Is China a Risk or an Opportunity for Europe?

Labour, Education & Society

Edited by György Széll, Heinz Sünker, Anne Inga Hilsen and Francesco Garibaldo

Volume 11

PETER LANG
Frankfurt am Main · Berlin · Bern · Bruxelles · New York · Oxford · Wien

Francesco Garibaldo
Philippe Morvannou
Jochen Tholen
(eds.)

Is China a Risk or an Opportunity for Europe?

An Assessment of the Automobile, Steel, and Shipbuilding Sectors

PETER LANG
Internationaler Verlag der Wissenschaften

Bibliographic Information published by the Deutsche Nationalbibliothek
The Deutsche Nationalbibliothek lists this publication in the Deutsche Nationalbibliografie; detailed bibliographic data is available in the internet at <http://www.d-nb.de>.

The research project on which this book is based was commissioned by the European Metal Workers' Federation (EMF). The EMF is grateful to all the people who contributed to the study and who generously shared their knowledge about China. The EMF would also like to thank the researchers of Syndex (Paris,) the University of Bremen and the Istituto de Lavoro (Bologna), and the European Commission for funding of the project. Sole responsibility for this book lies with the authors; the European Commission cannot be hold responsible for any use that may be made of the information therein.

ISSN 1861-647X
ISBN 978-3-631- 57474-4

www.peterlang.de

Table of Contents

Acknowledgement 7

Foreword 9
Francesco Garibaldo, Philippe Morvannou, Jochen Tholen:

Chapter 1
Is China's growth sustainable? 15
Emmanuel Reich

Chapter 2
Relations between China and the European Commission 23
Caroline Jacobsson

Chapter 3
The automotive Industry 27
Francesco Garibaldo, Oscar Marchisio, Volker Telljohann

Chapter 4
Shipbuilding in China and its impacts on European shipbuilding industry 53
Thorsten Ludwig, Jochen Tholen

Chapter 5
Steel 87
Daniela Gradinaru, Philippe Morvannou

Chapter 6
Policy recommendations 111
Caroline Jacobsson

Annexes 113

References 117

Abbreviations and Vocabulary 121

Authors 125

Acknowledgement

The research on which most of this book is based was funded by the European Commission and managed by the Secretariat of the European Metalworkers Federation EMF. We should like to express our appreciation to the EU Commission and the EMF Secretariat for enabling us to collaborate.

The research had been carried out by using the all references/sources and doing interviews with experts all over Europe. Though, due to financial restrictions, the teams could not do interviews in China by themselves, their partners have done empirical research in China itself and delivered the material to the authors. There are dozens of individuals – the interviewees and the respective partners – whom we need to thank for their assistance and transfer of their knowledge.

Foreword

Francesco Garibaldo, Philippe Morvannou, Jochen Tholen

1 Background

China's industrial growth has been the unexpected event of the beginning of the 21st Century owing to its impact on the entire global economy. This has led the EMF into discussions about the impact of the developments in China on the European metal sectors. The EMF China Project consists of research on the future of European metal industries. The aim is to get a common understanding of China's economic future and its consequences for Europe based on the situation in three metal sectors chosen as examples, namely steel, shipbuilding and automotive sectors. That assessment can then serve as a basis for describing and evaluating the importance of the factors enabling Chinese industry to be globally competitive. Finally, this research will endeavour to pinpoint the relevant issues and establish ways of internationally regulating the growth of China's metal industries, making them compatible with international standards regarding sustainable development and social responsibility.

During the beginning of the 21st Century China has become the world's fourth largest exporter and is the third largest importer, and the recent scrapping of quotas in the textile and clothing sectors will merely strengthen its position. China has become the reference market for ferrous and non-ferrous metals alike, if not for raw materials in general. China has become one of the world's main creditors involved in financing US debt and awarding loans to the developing countries. In short, China's recent emergence onto the world economic scene is shaking up long-established hierarchies and raises the question of the future of its relations with the world's major economic areas, including Europe.

If one sector of the economy has been at the heart of China's breakthrough it is that of the manufacturing industry. This is not because China has become industrialised in recent years, but mainly because it has transformed its industry over that period by competing on global markets.

One of the main reasons behind this transformation is the combined effect of investments by multinationals – including European multinationals – and the assertive industrial policy adopted by the Chinese authorities. A concomitant factor is the development of multinationals on the back of Chinese capital.

Following an initial phase of satisfaction the domestic demand, exports based on comparative advantages are expanding to increasingly elaborate products that are gradually starting to compete with European markets both within Europe and in third markets. So relocation is going hand in hand with the loss of external markets. A new international division of labour is occurring before our very eyes, and it is playing a part in weakening European industries in the long-term.

However in spite of China's accession to the WTO, this danger -which is posing more and more of a threat to European industry -is based on unsustainable competitive advantages. This is because the rules governing Chinese society and freedom of association do not meet international standards; environmental impact is not controlled to a great extent - if at all; and more often than not intellectual property rights are simply ignored. So it is the European model of development and its social dimension that are likely to be called into question by a form of competition that is essentially unfair.

What impact on Europe's manufacturing industries is China's eruption onto markets for manufactured goods having and what potential impact could it have? Will China's expansion continue? If so when and under what conditions? How much leeway does Europe have in the area of industrial policy? To what extent does the Lisbon Strategy, which is seen as a balanced objective combining economic growth, social progress and sustainable development, present a coherent European response? What consequences is China's impact having on the strategies pursued by European companies, especially relocation and its positions in emerging markets?

More importantly, does the 'Chinese industrial revolution' not call for innovative responses -not just industrial innovation but also innovation in companies, in integrating lifelong learning into the promotion of skilled jobs - and the organisation at international level of structured social dialogue between China and Europe so as to highlight the value added of the European Social Model?

2 The Project

The research on which most of this book is based was funded by the European Commission and managed by the Secretariat of the European Metalworkers Federation EMF and carried out in 2006.

The aims of the project were to

- Draw up a shared assessment of the immediate future of the European industries in question
- Initiate a discussion on the role of the social actors in regulating the European industries in question via social dialogue bodies like European Sectoral Social Dialogue Committees, European Works Councils, international dialogue committees within multinationals and regional social dialogue committees
- Hold a seminar to discuss lines of action to be taken by the trade union actors vis-à-vis multinationals and the actors behind China's industrial development.

The project has been conducted by one research group and one Steering Committee.The research group consists of three research teams and a project leader from the EMF Secretariat. Each team was responsible for one sector and one of the teams was also coordinating the research work as a whole.

Research on the steel sector was conducted by Philippe Morvannou and Daniela Gradinaru at Syndex in Paris, which was also coordinating all three research teams. Research on the shipbuilding sector was conducted by Dr. Jochen Tholen and Thorsten Ludwig at the University of Bremen. Research on the automotive sector was conducted by Francesco Garibaldo at the Instituto per il Lavoro in Bologna.

The three teams Syndex in France, the University of Bremen in Germany and the Instituto per il Lavore from Italy evaluated the current stage and further developments of three branches of Chinese metal sector (steel, shipbuilding and automotive), with a strong reference to the interdependencies with the respective branches in Europe.

The research had been carried out mainly in 2006, by using the all references/sources and doing interviews with experts all over Europe. Though, due to financial restrictions, the teams could not do interviews in China by themselves, their partners have done empirical research in China itself and delivered the material to the authors.

The Steering Committee consisted of representatives from the EMF Industrial Policy Committee, the EMF Company Policy Committee and representatives from each of the steel, shipbuilding and automotive sectors (in total nine people).

There has also been cooperation with the International Metalworkers' Federation (IMF) and participation in the IMF China Work Group in Geneva.

3 The content

There have been several streams of output, to be presented here.

Emmanuel Reich starts with the question, whether China's economic growth is sustainable; a question which answering is not only important for global capital, but for the Chinese people themselves and for the future of European business. Sine 2003 till 2007 there is an annual growth rate of the Chinese economy of about 10 percent, though the dimensions should be considered: The Chinese GDP per capita is just one third of the Polish GDP per capita or roughly 5 percent of the respective US figures in 2005. Though there are many uncertainties for a linear growth (aging population, lack of democracy, climate change and ecological crisis) and though nobody honestly can do a prognosis on a 75 years basis, there is much room for China as a still developing country to catch up with the industrialized world.

Caroline Jacobsson is presenting the official relations between China and the European Commission and describes, which kind of channels between China and the EU are available for information, assistance and help. The EU has established its own Chamber of Commerce in China, the European Commission organises about 20 Dialogue Groups on sectoral level in Beijing, and there is a China Desk of the European Commission, mainly organized by the DG Trade and DG Relex (External Relations).

Francesco Garibaldo, Oscar Marchisio and Volker Telljohann start with the first part of the core of this research, the presentation of the automotive sector. In 2005 China became the second world market and the third world producer of cars. According to the new policies for the Chinese Automotive Industry from 21 May 2004, the Chinese industry is getting more sophisticating with the aim to become exporter of cars in the medium range.
Though there are some social institutional constraints and delocalization limitations, it could be said, that there are impacts of the Chinese automotive sector on the European one. Not only the FDI by the European carmakers in China (which impacts are largely well-known), but vice versa in the mid run by exporting Chinese cars to the European market as the home and reference market of the European OEM. But even here the challenges for the European OEM are not the same for all car producers. Insofar one recommendation is to maintain a strong specialisation such as the premium segment, the other is to move the European sector in the direction of a sustainable mobility (new generations of engines, new mobility patterns for urban areas, developing an ecological dismantling cycle. And overall: to develop the core of the European strength, the "R&D" sector.
Thorsten Ludwig and Jochen Tholen are continuing the sector presentation by studying shipbuilding industry. From 1975 till 2005 the market shares in global shipbuilding of the countries of the todays EU 25 went down from 38.9 percent to 6.9 percent. China's shipbuilding industry showed up on the world stage only in 1985 with only 0.9 percent, but in 2005 it had already achieved 13.8. percent (number one in 2005 was South Korea with 37.7 percent followed by Japan with 35.0 percent). The goal of the Chinese shipbuilding policy (which production is mainly concentrated on bulkers, tankers, container vessels) is to become in 2015 No 1 in the global shipbuilding market with 35 percent market share. Though - by presenting four small case studies of Chinese shipyards - identifying constraints of the continuous growth of the Chinese shipbuilding industry like the fields of labour relations, collective bargaining rights, wages, migrant workers and qualification, Health and Safety, social insurance etc., our assumption is that China's run for becoming the world champion in shipbuilding will be successful even five years earlier than officially planned. What could be the recommendation for the European shipbuilding industry? Intensifying the concentration on the segment of high sophisticated ships (amongst others cruise ships, RoRo ships, advanced ferries, but container ships as well), improvement of the vertical relations of the European shipbuilding industry with the marine supply/equipment industry, improvement of the (insufficient) horizontal cooperation between the yards themselves, reinforced cooperation also in sensitive areas (like R&D, qualification), development of a Pre-and-After-Sales service by the European shipbuilding industry. Promising approaches are the initiative "Leadership 2015" of the European shipbuilding industry as a basis for a joint

industrial policy and the EU Social Dialogue Shipbuilding, which is gathering the "stakeholders" of the industry on the European level.
Daniela Gradinaru and Philippe Morvannou are closing the sector studies by presenting the Chinese steel industry. Propelled by the enormous consumption of steel coming from all sectors of the fast growing industry in China, and gradual substitution of imports by domestic production, there was a remarkable growth between 2003 and 2005 (annually 20 percent and more) of the Chinese steel production (including pig-iron production and raw material production (iron ore and coke)). Since 2006 there is a slight slowdown of the steel production as a process of adaptation to the forecasted demand for steel in 2010 of 480 million tons. The renewed Chinese policy for the steel sector from 2005 means for the future the concentration of production and sale according to the needs of the Chinese economy. Traditionally (and different form the other sectors presented) the share of FDI in the Chinese steel industry is pretty low (only 1.4 percent of the overall FDI in China), and there is no policy to export Chinese steel to the global markets. But despite of that, the effects on the global and here the European steel industry are enormous and profitable, because of the chances of export for European steel and the very high prices on the world markets. On the other hand, there is now a run for raw materials (mainly iron ore, coal and scraps) and the dramatic rise of the steel prices will have effects not foreseeable now on other steel-consuming sectors worldwide. Furthermore there is a fear that the development of the Chinese steel producing capacities could lead to global overcapacities and a need for the European commission to put this question on the agenda of the "China Desk".
Last but not least *Philippe Morvannou* is concluding with some general, mostly supra-sectoral policy recommendations. Starting with the insufficient labour relations in China, stronger ties between companies TU representatives in China and European TU activists (either working in European MNC having FDI in China or more generally involved in projects dealing with China) are requested to create networks. And a more closer look should be taken to the Chinese activities in Europe, which could not only be a threat to the European industry, but could offer opportunities as well.
Furthermore attention should not only given to China, but to the whole region (like Vietnam) and other world regions knocking on the door of the world economy (like India, Brazil etc.).
On the other hand we should consider those countries which are direct competitors of China and could threaten Chinese growth. This means not only Europe, but especially Japan, Korea etc.
Last but not least the European Commission is requested to act more directy, not only in the question of shaping the labour relations in China according to international and ILO standards, but in terms of a coherent European industrial policy for the sectors. An example could be the creating of a European Maritime Coordinator.

Chapter 1
Is China's growth sustainable?

Emmanuel Reich

1 China's emergence in the global Economy

1.1 China's growth has been striking

An impressive growth rate
Since the last 20 years, China has recorded more than 9% per year of GDP growth rate and this trend does not seem to decelerate. China is even growing faster since 2004-2005 despite measures taken by the authorities.

Fig. 1 - GDP annual growth source OECD

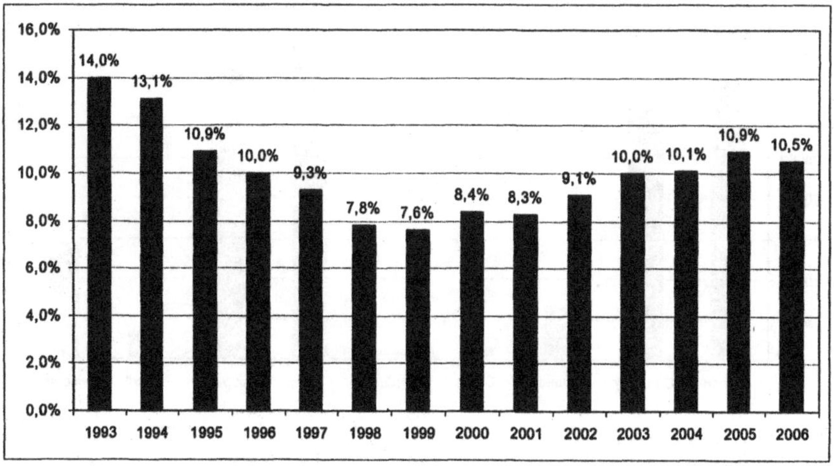

Source: OECD

The new assessment of China's growth rate and GDP gives a more balanced view of its economy.The new figures give a better estimate of services in the Chinese economy, of the investment rate and the export/import rate.The export and import rates show China's high reliance on external markets.These new figures imply that China's GDP is almost as important as France (5th world economy) and that it represents 5% of world GDP instead of 4%. In 2020, it could account for more than 20% of world GDP if growth continues at a minimum of 7% per annum.

GDP per capita and reduction of poverty

One of the positive aspects of China's growth has been the increase of its GDP and GDP per capita. Although this criteria is narrow minded (for instance the UNCTAD Human Development Index takes into account not only GDP which is solely a monetary criteria but also health, life expectancy and literacy), it shows China's performance in the last 20 years. In 25 years, average revenue has been six times higher than before and people living below the poverty line have been divided by ten. Poverty eradication has been impressive in the 1980's though it stagnated in the 1990's.

The table shows that GDP per capita in China is still very low. Compared to India, China has a clear superiority; these are the benefits of early reforms. But compared to developed countries or even emerging ones, China is still far behind.

Fig. 2 - GDP per capita $

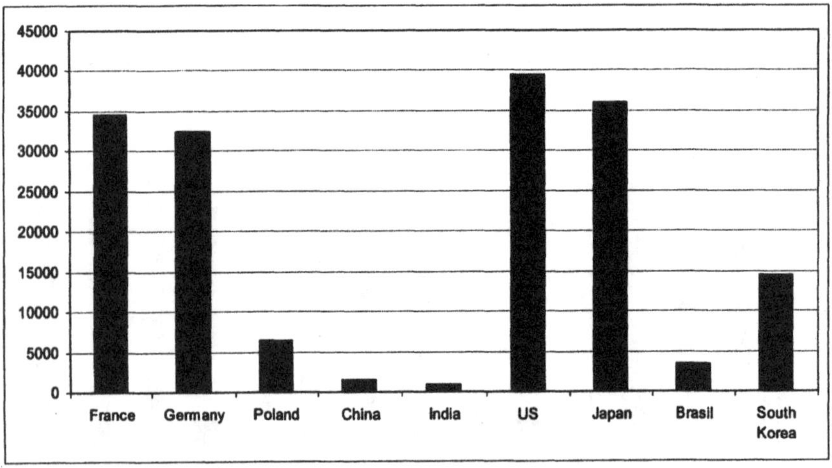

Source: WIR 2005 UNCTAD

1.2 FDI flow to China

FDI flow to China has increased during the last years making China one of the most popular destinations for FDI in the world. Two thirds of the FDI come mainly from Japan, Korea and Hong Kong.

As a result of the FDI flow, foreign companies do play a major role in China's productivity and exports (57% and even 80% in the high tech products).

Flows from EU25 are much more modest. EU FDI are mainly internal flows (more than two thirds) the rest of the FDI go to the US, Japan, Canada and China. Only a small fraction of European investments are turned to China.

Tab. 1 - Number of greenfield FDI projects 2002-2004

	2002	2003	2004
World	6.556	9.303	976
EU	1.770	2.565	2.851
United States	414	599	578
China	**581**	**1.299**	**1.529**
India	250	457	695

UNCTAD WIR 2005

Tab. 2 - FDI flows (2005)

	Export to:						Import from:					
	Intra EU25	Extra EU25	Of which				Intra EU25	Extra EU25	Of which			
			US	CH	JP	CA			US	CH	JP	CA
EU 25	354,5	152,5	26,1	5,9	-1,2	9,3	324,4	69,8	17,5	7,3	4,2	6,9

1.3 Exports as a main driver

Tab. 3 - Millions of economically active persons

	Global	Advanced countries	Least developed countries	New
1980	960	370	590	
2000 before	1.460	460	1.000	
2000 after	2.930	460	1.000	1.470

New: China 760; India 440; Ex-soviet 260; Others 10
Source: ILO and Richard B. Freeman

China's rise in the international trade has made the country a key actor. In the last two decades, China's share in international trade jumped from 2% to 7%. China has become the third largest exporter in the world and is expected to become the largest at the end of the decade. China is already enjoying huge trade surpluses especially with the United States. On the other hand, one should keep in mind China's high dependency on other Asian economies. China has a substantial trade deficit with South Korea and Japan, which export components, semiconductors, chips which are assembled and exported by China to Europe and North America.

1.4 Doubling the global workforcel[1]

In 1985 the population of North America,Western Europe, Japan, the Asian Tigers and Africa consisted of 2.5 billion people. Since China's shift to capitalism, India's end of autarchy and the end of Soviet communism, the world consists of more than 6 billion people. For the work force, figures evolved from 1 billion in 1980 to 3 billion in 2000 including the newcomers China, India and

[1] *Doubling the Global Work Force: the challenge of integrating China, India and the former Soviet Bloc into the World Economy?*, Richard B. Freeman, November 2004

Russia.The doubling of the global work force has created a massive competition.This is not contradictory with the existence of segmentation on the Chine labour market. Differences in skills and wages among regions and provinces (i.e. East coast vs.Western China) are quite severe.

2 Is China's growth sustainable?

2.1 Major economist disagree on China's future

The Wall Street Journal has confronted twelve major economists on the prospects of China's economy.The economists are almost unanimous in saying that the Chinese boom will continue for a long time. Some of them believe that China will become the largest economy in the world in 75 years. This prognosis is referring to GDP in value and not GDP per capita.
Akerlof believes that Europe will become the biggest economic zone because of its large population. But the gap per capita between the US and Europe on one hand and China on the other is too large. Arrow thinks that if China keeps growing like it has, it will become the biggest economic power (in value). Coase and Markowitzdo agree on this idea as well as Friedman. Granger still believes the US will be the largest economy followed by Europe and China. Smith agrees on this prognosis.
Stieglitz considers that it is difficult to make a prognosis on a 75 years basis. One does not know how an enlarged Europe will perform and in what manner it will enlarge.
Another hot issue is whether the Chinese population will continue to grow. Even if China does not grow as fast as 10% per year, the gap will decrease and a new equilibrium will emerge on the world economic scene.

2.2 Factors behind growth are there for a while

GDP is the product of hourly labour productivity, average number of working hours and total employment (determined by the participation and unemployment rates).

Therefore, Gross Domestic Product (GDP) can be raised through the following factors:
- Productivity
- Number of working hours
- Quantity of labour

Average productivity is seven times higher in the industry sector compared to agriculture. Productivity in the services is three to four times higher.
It means that the millions of rural workers that immigrate to the cities will mechanically contribute to an increase of the GDP in the coming years. 120 million rural workers are currently living in cities.This figure is likely to reach 300

million by 2020. Moreover, huge efforts are made in education.The average level of education and skills is rising quickly.
Based on assessment[2] (labour and wage growth estimated through productivity), simple calculations show that the Chinese growth should have been around 5-6% per year (more than 9,5% in reality). This means that if 1% of the labour force shifts from agriculture to industry, it adds 4 to 5% GDP growth rate per year. For the 2005-2015 period the predicted growth is around 10%.

2.3 Negative external factors could hinder China's growth

Violation of labour rights and ILO rules have made China one of the least developed regimes for workers and unionists
ILO (International Labour Organisation) reports have highlighted the violation of ILO basic standards. The ICFTU (International Confederation of Free Trade Unions) has reported this as well. Some attempts have been made to create new trade unions but the authorities have always reacted strongly against this and sent trade union leaders and activists to jail.
As a reaction to exploitation, corruption and social unease, social unrest have increased since the 1990's. In 2004 and 2005 more than 70.000 demonstrations took place compared to roughly 10.000 demonstrations ten years ago.

Ecological crisis, pollution and climatic change
Because of the communist regime and the absence of counter-powers, the economic development in China suffers severe weaknesses. Pollution is degrading the quality of water and much of it has become undrinkable. 60% of the water in the rivers cannot be used for consumption. Water is becoming a hot issue: natural water available for the inhabitants is insufficient.The gap between availability and needs is increasing, creating a cut in the supply.
Chinese cities are amongst the most polluted of the world. 16 out of the 20 most polluted cities of the world are Chinese. Erosion and desertification are growing fast, approaching the suburbs of Beijing. China uses coal abundantly. Two thirds of its energy needs is filled by coal, which is very polluting.
Quite often land expropriation has been conducted without compensation to the inhabitants. Big dams are built forcing thousands of people to leave their villages and funds set aside to rebuild the villages in other part of the area are embezzled by officials. The damage on the environment has its cost. People loose time in traffic jams and traffic accidents are common. Recycling water is costly. Erosion and desertification limit and threaten crops and agriculture. The financial losses caused by pollution may account for as much as 10% of GDP according to official estimates.
The US development model cannot be enlarged to the whole planet. If it could several planets would be needed to supply all the natural resources needed to

---

[2] *Why China's rise is sustainable*, Carsten A. Holz, "Far Eastern Economic Review"

sustain the high level of pollution.The world and China will have to change their model of development in order to comply with environmental issues.

Unemployment, migration and inequality are hot issues

China has a a high level of unemployment. Millions of workers were sacked because of the restructuring of public companies. Furthermore every year millions of rural labour migrants enter the cities and fuel unemployment.

Between 1990 and 2004[3], China has created 7 millions jobs per year but active population was increased by 10 million.Industry is the most dynamic sector but it creates few jobs. In the coming years, China should create more than 300 million jobs to avoid an unbearable level of redundancy.

China has become one of the most unequal countries of the world.That is an amazing and sad achievement.The World Health Organization (WTO) has classified China's health system as number 181 out of 191. Inequalities of revenues between rural and urban regions could be a ratio of one to six.

Demography: China will be old before it becomes rich

The one-child-policy was implemented severely in the 1980's and is provoking a demographic shock. There are 115 boys for every 100 girls.Ageing of the population will come quickly and the working population will be reduced after 2015.

At this time no public health system exists nor any pension system.The Chinese population is ageing and 10% of the population is over 60 years old. In 2030 this figure will be 30%.The need for healthcare will rise.This situation will be a heavy weight for the country. Furthermore, the stabilization of the population will limit the potential growth of the country in the future.

Other economic factors such as bankruptcy of the banking system or overcapacities in some sectors, for instance in the steel or automotive sectors, could produce sporadic crisis.

2.4 Towards a smooth landing?

An economic policy orientated to household consumption?

A reorientation of the economic policy to household consumption could be implemented by the development of public services in rural regions. The rise of wages would allow more domestic consumption of cars and electronic goods as well as of services.A local industry that turned to recycling and sustainable energy would also create a many jobs and contribute to a better environment.These reforms would make China less dependant on world trade.

For the moment only sustainable development has been endorsed by the Chinese government, which took it into account the need to reorientate its economic development in its XI Plan. But this slight reorientation is not enough considering the issues at stake.

---

[3] *Chine: la croissance et ses déséquilibres*, document de travail de l'AFD. Janvier 2006.

Democratization can bring a lot to China

Beyond basic human rights, democratization could also produce counter powers to limit the reign of arbitrary and corruption and enhance the rights of the individuals. Free trade unions, labour standards and increased wages would contribute to China's domestic consumption as well as produce a better equilibrium with the rest of the world.

Chapter 2
Relations between China and the European Commission

Caroline Jacobsson

1 The European Union Chamber of Commerce in China

The European Chamber in China was officially established in Beijing in October 2000 under the initiative of the European Union Delegation and with the financial support of 51 European companies.
The European Chamber is a non-profit organisation supported by members and Corporate Sponsors and has at present offices in Beijing, Chengdu, Guangzhou, Nanjing, Shanghai and Tianjin. Plans are also in place to develop a European based office in Brussels during 2006.

1.1 Objectives

The European Chamber is the main private European lobbying organisation in China, representing the interest of European companies present or active in China. Its main goals are:

- To establish a constructive dialogue with the Chinese authorities in order to improve the business environment in China
- To monitor and report on market access in China and present concerns and recommendations in the Position Paper
- To establish effective communication with the EU Delegation, the European Commission and Member States
- To host functions that allow our members to discuss and network with Chinese and European Officials

1.2 Working Groups/Lobby Actions

With lobbying as the main business scope, the main internal organisation of the European Chamber is focused on assisting industrial sectors present in China according to the needs of members. At present there are 27 sectoral and horizontal Working Groups, as follows:

- Aerospace
- Automotive
- Automotive Components
- Aviation
- Banking
- Construction
- Cosmetic
- Corporate Social Responsibility
- Energy and Utility Policy
- Finance & Taxation
- Food and Beverage

- Human Resources
- Information Technology and Telecommunications
- Insurance & Fund Management
- Intellectual Property Rights
- Legal
- Logistics
- Maritime Transport
- Operations
- Petrochemicals, Chemicals, Oil and Gas
- Pharmaceuticals
- Public Procurement
- SME
- Sourcing
- Trade and Distribution
- Travel
- Wine & Spirits

Each Working Group contributes to the annual Position Paper on European Business in China. The Position paper is presented annually to Chinese authorities as well as the Foreign and Chinese press from the beginning of September. http://www.europeanchamber.com.cn/

2 The European Commission in Beijing

The European Commission in Beijing organises several Dialogue Groups on sectoral level.There are about 20 groups, for instance steel, trade, cars, ship, transport, environment and energy.The dialogue groups work through a mainly oral exchange of news. There is no common approach or understanding between the parties in the dialogue. Sometimes documents are exchanged, but only documents of official nature. The dialogues are bilateral, where one party is the European Commission and the other party is a Chinese ministry in the sector which is the target of the dialogue (for instance the ministry of trade or ministry of energy).There are 1-2 meetings per year.They take place in China and the languages vary depending on who participate so it could be Mandarin Chinese or English with interpreters. 3-4 people and up to 10 people from each side participate. The construction of each meeting is ad hoc and the purpose of the dialogue is to lead to a joint initiative, but not more than that. There is no "master plan" as Mr. Franz Jessen, Deputy Head of the Delegation, puts it.The dialogues have been going on for about 4-5 years (since the beginning of 2000) and they are always bilateral on a high level. Companies or other parties are never involved.
One concrete example of a bilateral dialogue group is the one concerning the export of banned pharmaceutical products to Europe. Instead of letting the ex-

port proceed the dialogue informed the Chinese that the products would be banned by a new law and therefore no effort to export them took place.
http://www.delchn.cec.eu.int/

3 The European Commission in Brussels

3.1 DG Trade

The dialogue between the European Commission in Brussels and China is held on four levels. A summit is held on national level between heads of state (level 1). Joint committees are held on minister level. These are meetings on different subjects between different ministers (level 2). Meetings are also held on DG and vice minister level (level 3) and low level meetings are held between the Ministry of Commerce (MOFCOM) and units at DG Trade (level 4).
One example of a level four meeting is the specific sector dialogue steel group called the "steel contact group".This group meets once or twice per year. These meetings include also the industry itself and handle news related topics. The information flow during these meetings go both ways and the purpose of the meetings are in general maintaining trade relations.

3.2 China Desk

The China Desk is bilateral and handles overall strategic work, coordination, preparation and arrangement of meetings between the two parties.There is co-operation between the China Desk at DG Trade and specific sectors. Like steel or auto.All sectors are handled by the China Desk all dependant on what is needed at the time.The sectors are therefore problem oriented and provide support to the China Desk when meetings are called.The China Desk was founded in the late 1990s and the overall goal is to integrate China as a partner in world economy. The European Commission wants to influence the way China is thinking, for instance in adopting standards or legislation that improves the conditions for cooperation in the area of trade.
Trade union issues are per say not included in the work of DG Trade, but under the China section under DG Relex (External Relations).
The difference between DG Trade and DG Relex can be explained like this: DG Trade DG Relex Geographic desk: China Geographic desk: China Topic:Trade Topic: Everything but trade
http://ec.europa.eu/comm/trade/
http://english.mofcom.gov.cn/
http://ec.europa.eu/comm/external_relations/index.htm

Chapter 3
The automotive industry

Francesco Garibaldo, Oscar Marchisio, Volker Telljohann

1 Auto sector development in China

1.1 Overview of the market and the main producers

The turning point of the Chinese development in the auto sector came in 2001, date of the admission of China in the World Trade Organization (WTO). Before there was an initial period (1953-1983) of low production capacity and outdated technology, followed by a first wave of modernization (1984-2000) during which the Chinese government proposed a 'market for technology' strategy, which aimed to get foreign manufacturers to transfer their technology to Chinese companies so as to allow them to win back market share from the foreign players.

The first Sino-foreign automobile joint venture was formed in 1984 by DaimlerChrysler and Beijing Automotive Industry Corp. Later on, other global auto manufacturers, such as General Motors,Volkswagen, Ford, Toyota, Nissan-Renault, PSA Peugeot Citroen, Honda and BMW also established joint ventures in China.

The establishment of joint ventures helped China's automotive industry develop at a faster rate mainly because of the huge transfer of technology and managerial techniques from Western countries and Japan. However, the joint ventures controlled about 90% of China's passenger car market at that time and have dominated the market since.

After China entered the WTO in 2001, the automotive industry began to grow even faster.The overall output increased by 38.83% and 36.69% in 2002 and 2003, and the output of sedans were up by 55.2% and 85% respectively. During these two years, there were almost zero inventories at year end. Then there was a slowdown in 2004 because the Chinese government launched a few finance-based polices to cool down the market, such as discouraging bank lending and slowing the approval procedure for investments. In addition to these macro controls, lower personal car loans from banks and frequent prices cuts also kept consumers away. Many potential buyers delayed their purchases as they expected prices to drop further.As a result, production and sales of automobiles started slowing down after the booming years of 2002 and 2003. In 2004, the total output and sales (excluding imports) of automobiles climbed by 14.2% and 15.5% to 5.074 million and 5.071 million units respectively. Sedan output and sales for automobiles came up to 2.31 million units and 2.33 million units respectively.

In 2005, China's automotive market continued to maintain stable growth, with total output reaching 5.71 million units and auto sales hitting 5.76 million units, a growth of 12.55% and 13.54% respectively year on year. By adding about 160,000 units of imported automobiles in the year, China's automobiles sales came close to 5.92 million units.

2005: China becomes the second world market and the third world producer

In the year of 2005, China surpassed for the first time Japan's sales of 5.8 million units to rank as the world's second-largest automobile market in terms of numbers of cars sold, only next to the U.S. Apart from being a huge market for cars, China has also become a major producer. In 2005, China became the world's third-largest producer, ahead of Germany, only second to the U.S. and Japan.Without a doubt, China has become a large automobile consumer and producer market.

However, the Chinese auto market is far from saturated according to the economic size, strength and growth model of the country. For example, the total highway mileage of China is ranked second in the world; however, the ratio of the total number of vehicles to the total highway mileage is only one-third of the U.S., one-fifth of Japan, one-sixth of Germany and half of South Korea. Besides, there are currently 24 units for every 1,000 people in China, still far behind the figure in the U.S. (700 units per 1,000 persons), and the global average, which are 120 units per 1,000 persons[1]. In recent years, the fast-expanding passenger car market acted as the driving force for the booming automotive industry because Chinese consumers had more auto models to choose from as more automakers entered the market after China joined the WTO. Also, the increasingly growing disposable income of the Chinese people has boosted demand for cars. Unlike a few years ago, when automobiles were mainly purchased by the government and corporations for business use, they have now become personal consumables and private car consumption has gone mainstream. By the end of 2005, autos purchases for individual use has exceeded 50% of total sales and more than 70% of purchases are in the urban areas.The possession of private vehicles has exceeded 23 million units in 2005.

On the other hand, the gap also provides much opportunity for the development of the auto industry. Assuming one in every 100 persons purchases one automobile in a year, the demand for automobiles in the country will be 13 million units, more than double the country's auto production in 2005. Although the GDP per capita of China is only US$1,700 so far, the average GDP per capita of the coastal region has reached about US$4,000.The purchasing power of the Chinese people will lead to a rise in demand for household cars.Therefore, the existing vehicle reserve of China is still very small and the market potential is huge, although the auto output and sales have been growing very fast.

1Source: China Association of Automobile Manufacturers

The development of China's auto industry is expected to develop at a mature and stable rate. Driven by the country's steady economic growth, the auto mar-

---

[1] Source: China Association of Automobile Manufacturers

ket will keep expanding and is widely expected to exceed 10 million units by 2010 and 16 million units by 2020.

Changes in the industry

In the past five years, there have been a few positive changes in China's automotive industry:

- The Chinese government reformed the automotive industry policy and relaxed its control over the industry.
- The government began to encourage and support private auto consumption which helped expand the passenger car market.
- Along with the increase of foreign investments and the entry of private capital into the industry, the overall production capacity and the economy of scale have been growing quickly.
- As the barrier of entry was lowered, the market environment became more competitive and this drove vehicle prices down.
- The auto parts industry grew quickly, and with it, specialization and globalization.
- Automobile-related services, such as auto finance, repair, maintenance and insurance started progressing.
- The government sped up the construction of transportation infrastructure to support the growth of private automobile ownership.

Tab. 4 - Statistics of automobile output in the past 10 years

Year	Sedan		Coach		Truck		Total	
	Output (million	Growth rate	Output (million	Growth rate	Output (million	Growth rate	Output (million	Growth rate
1996	0.3829	--	0.1895	--	0.6251	--	1.4752	--
1997	0.486	26.93%	0.2656	40.16%	0.5736	-8.24%	1.582	7.27%
1998	0.5071	4.34%	0.3211	20.9%	0.7356	28.24%	1.63	3%
1999	0.571	12.60%	0.4249	32.33%	0.8396	14.14%	1.832	12.39%
2000	0.607	6.3%	0.5779	36.01%	0.8629	2.78%	2.07	12.99%
2001	0.7036	15.91%	0.7207	24.71%	0.8901	.15%	2.3417	13.13%
2002	1.092	55.2%	0.8647	19.98%	1.092	22.68%	3.251	38.83%
2003	2.0201	85%	0.9466	15.39%	1.1244	2.97%	4.4439	36.69%
2004	2.314	11.76	1.1584	24.23%	1.5849	40.96%	5.0741	14.2%
2005	2.77	24.42%	--	--	--	--	5.707	12.56%

Source: China Statistic Yearbook

Foreign Direct Investment (FDI)

Apart from transfer of technology the main drivers of the new wave of development in China were the huge investments in Joint Ventures by USA, European and Japanese OEMs (original equipment manufactures, that is the assemblers of cars). In 2002 and 2003, the thriving auto market attracted a lot of local and foreign investments that built or expanded the capacity of production plants in China. In 2002 and 2003, foreign automakers invested a total of US$6.3 billion in China. However, the increasingly fierce competition in the market led to lower profit margins for all the companies. This prompted the government to start controlling the overheated investments in 2004.

Automotive industry in 2005

Before 2005, China classified automobiles into mainly three categories -truck, coach and sedan. In January 2005 the China Association of Automobile Manufacturers (CAAM) released a new categorization standard based on car model. Under the new classification, automobiles were classified into passenger cars and commercial vehicles.

The passenger car is used for the transport of passengers, and their accompanied baggage, comprising no more than nine seats including the driver's seat, and can be linked to a trailer. Passenger cars include sedans, MPVs (multi-purpose vehicles), SUVs (sport utility vehicles) and crossover vehicles (such as mini-vans).

The commercial vehicle is used for the transport of passengers and goods, comprising more than nine seats including the driver's seat. Commercial vehicles include coaches that are nine seats and above, trucks, semi-trailer towing vehicles, incomplete coach vehicles and incomplete truck vehicles.

According to the statistic from CAAM, by the end of 2005, China has in total 145 car manufacturers, an increase of 16 over 2004.

Tab. 5 - Number of car manufacturers in China

	2003	2004	2005
Car manufacturer	102	129	145

Source: China Association of Automobile Manufacturers

In 2005, China's automobile market rebounded from the stagnant market of 2004. Average monthly sales volume surpassed 475,000 units and December was a record month as sales approached the 620,000-unit mark. The performance of the passenger car market improved, with output and sales up by 12.56% and 13.54% respectively. In 2005, sedans continued to be the major force behind the growth of the automobile industry, accounting for 49% of total automobile sales. In contrast with the booming passenger car market, the commercial vehicle segment declined by 0.6% and 0.75% in terms of production and sales respectively (Tab. 6).

Currently, the top five car makers in China are FAW Group, SAIC (Shanghai Automotive Industry) Group, Dongfeng Group, Chang'an Group and Beijing Auto Group. In 2005, the top five automotive groups sold 3.86 million autos, accounting for 67% of the total sales in China (Tab. 7). The sales for each company exceeded 500,000 units over the year. The car makers ranked sixth to tenth are: GAI (Guangzhou Automobile Industry) Group, Hafei Motor (Harbin Hafei Motor Co., Ltd), Chery Group, Jianghuai and Geely Group.Among the top 10 automakers, Chery and Geely are the only two companies that are joint ventures. In 2005, the top 10 car makers sold 4.82 million units, accounting for 83.7% of the total sales in China (Tab. 8).

Tab. 6 - 2005 Automobile production and sales in China

	Production	Growth rate	Sales	Growth rate
Automobiles	5,707.7	12.56%	5,758.2	13.54%
Passenger Cars	3,930.7	19.73%	3,971.1	21.4%
Sedan	2,767.7	24.42%	2,787.4	24.31%
MPV	155.1	30.42%	155.8	42.76%
SUV	195.3	15.48%	196.4	20.63%
Crossover	812.6	5.47%	831.5	9.87%
Commercial Vehicles	1,777	-0.6%	1,787.1	-0.75%
Coach	176.8	-2.6%	178.6	-2.34%
Truck	1,162.4	4.26%	1,163.4	3.79%
Semi-trailer towing vehicle	56.5	-41.67%	56.8	-42.76%
Incomplete vehicle of coach	90.2	0.04%	90.5	-0.15%
Incomplete vehicle of truck	291	-4.34%	297.7	-2.98%

Source: China Association of Automobile Manufacturers

Tab. 7 2005 Top 10 automobile manufacturers in China (by sales, thousands of units)

Rank	Companies	2005	2004	Growth rate
1	FAW Group	983	1,013	-3%
2	SAIC Group	918	849	8%
3	Dongfeng Group	729	503	44.93%
4	Chang'an Group	631	505	24.95%
5	Beijing Auto Group	597	531	12.43%
6	Guangdong Auto Group	237	209	13.17%
7	Hafel Motor	230	205	12.16%
8	Chery Group	189	87	118.51%
9	Jianghuai	154	130	18%
10	Geely Holding Group	151	96	56.54%

Source: China Association of Automobile Manufacturers

Tab. 8 - 2005 Market share of top 10 auto manufacturers

Companies	Market share	Companies	Market share
FAW Group	17.1%	Guangdong Auto Group	4.1%
SAIC Group	15.9%	Hafel Motor	4%
Dongfeng Group	12.7%	Chery Group	3.3%
Chang'an Group	11%	Jianghuai	2.7%
Beijing Auto Group	10.3%	Geely Holding Group	2.6%
Total			**83.7%**

Source: China Association of Automobile Manufacturers

To some extent, the year 2005 was a turning point in China's automobile industry. First, the market structure underwent significant changes. Shanghai General Motors replaced Shanghai Volkswagen as the market leader in terms of sales.The market share of Volkswagen (combined with Shanghai-VW and FAW-VW) dropped to 17.9% in 2005. Also, demand for small-displacement and fuel-saving models went up in the mainstream market due to the impact of high oil prices. Most of all, Chinese self-owned brand auto makers have achieved impressive levels of improvement.The annual growth rate of Chinese self-owned brand car manufacturers stood at 53%, which is two times higher than that of joint venture companies. In the year 2005, domestic automakers occupied a 25% share of China's sedan market (Tab. 9).

Tab. 9 - Top 10 sedan manufacturers in China (by sales, thousands)

Rank	Company	Chinese Partner	Ownership	Foreign Partner	2005		2004		Share change
					Sales	Market share	Sales	Market share	
1	Shanghai GM	SAIC Group	State	General Motors	298.6	9.5%	252.1	10.9%	-1.4%
2	Shanghai VW	SAIC Group	State	Volkswagen	287.1	9.1%	355	15.3%	-6.2%
3	FAW VW	FAW Group	State	Volkswagen	277.	8.8%	300.1	12.9%	-4. 1%
4	Beijing Hyundai	Beijing Auto Group	State	Hyundai	224.7	7.2%	144.1	6.2%	1%
5	Guangshou Honda	Guangdong Auto Group	State	Honda	203.2	6.5%	202.1	8.7%	-2.2%
6	Tianjin FAW Xiali	FAW Group	State	Toyota	190	6.1%	130	5.6%	0.5%
7	Chery Auto	NON JV	Private	-	184	5.9%	86.6	3.7%	2.2%
8	Dongfeng Nissan	Dongfeng Group		Nissan	157.5	5%	60.8	2.6%	2.4%
9	Geely	NON JV	Private	-	149.9	4.8%	105.9	4.6%	0.2%
10	DPCA (Shenglong)	Dongfeng		Psa Peugeot Citroen	140.4	4.5%	89.1	3.9%	0.6%
Total					**2,112**	**67.4%**	**1.725.8**	**74.4%**	**-7%**

Source: China Association of Automobile Manufacturers

Note 1: As China implemented the new categorization standard in 2005, the market share is calculated by dividing the auto maker's sales by the balance of passenger cars minus the number of crossover vehicles.

Note 2: The sales records of Shanghai VW and FAW VW are calculated based on the Aak method. The Aak method refers to the number of car units sold to consumers, in contrast to the STD method, which calculates the number of units sold to car dealers.

Note 3: The sales records of FAW VW include Audi. Note 4: In 2004, Geely, a Chinese car maker based in Zhejiang, produced about 100,000 passenger cars and sold more than 100,000 passenger cars. However, as the company did not submit a consolidated sales report to the CAAM, Geely was not listed in the top 10 list in 2004.

Without exception, the top three carmakers failed to avoid losing market share in 2005. Beijing-Hyundai continued its ascent to fourth spot, nipping at the heels of the top three. Guangzhou Honda and Tianjin FAW Xiali consolidated their positions at fifth and sixth spots.
If Guangzhou Honda and Beijing Hyundai were the two stars of the automotive industry in 2004, then Dongfeng-Nissan and Chery were their equivalent in 2005. Most surprising was Dongfeng-Nissan, which saw its sales growing by 160% to 157,500 units in 2005; its profit margin was ranked first compared to all the other Chinese automakers. Sales for Chery rose 112% to 184,000 units, and its market share increased to 6.6% in 2005 from 2.9% the year before.
In 2004, only seven out of the top 10 automakers sold more than 100,000 automobiles. In contrast, the sales of all top 10 automakers in 2005 exceed 100,000 units.The top 10 auto makers occupied 73.3% market share in 2005.
In terms of sedan models sold in the market in 2005, Xiali ranked first in terms of sales in the sedan market.The small displacement models occupied half of the top list, and became the big winners of the year.The 10 most popular sedan models sold 1.26 million units in China, accounting for 45.53% of the total sales of sedans.

Tab. 10 - 2005 Top 10 sedan models in China (by sales)

Ranking	Model	Car Maker	Sales (thousands)
1	Xiali	Tianjilin FAW Xtall	179.7
2	Elantra	Beijing Hyundai	176.6
3	Santana	Shanghai VW	151.1
4	Excelle	Shanghai GM	150.9
5	Jetta	FAW VW	142.2
6	QQ	Chery Auto	116
7	Accord	Guangzhou Honda	114
8	Fit Saloon	Guangzhou Honda	89.2
9	Geely MelRi	Holding Group	75.6
10	Qtyun	Chery Auto	68
Total			**1,263.3**

Source: China Association of Automobile Manufacturers

1.2 New Automotive Industry Policy

The State Development and Reform Commission issued the Policies for the Automobile Industry on May 21, 2004. The Policies, which entered into force on May 21, replaced the Policies for the Automobile Industry issued in 1994.The Policies aim to restructure and strengthen the automobile industry in China and also provide guidelines for foreign investment in the industry.

Reforming the automobile industry

The Policies call for an overhaul of the Chinese automobile industry by the year 2010 in order to create large companies that can compete on the international market.As part of this overhaul, car and motorcycle manufacturing enterprises are encouraged to develop international cooperation and large enterprise groups are encouraged to associate or merge with international automobile groups.

Trademarks

Commencing from the year 2005, all cars and assembly parts manufactured in China will be required to indicate the registered product trademark of the manufacturer.All cars sold in China will have to display the trademark and name of the manufacturer as well as the place of origin in a prominent position.

Brand sales

The Policies require domestic and foreign car companies to establish a car brand sales and service system as soon as possible.This system can be set up by the manufacturer directly or by car distributors appointed by it. Both domestic and foreign investors will be allowed to conduct brand sales and after-sales service for domestic and imported cars provided they have been appointed by the car manufacturers and have completed the required formalities in accordance with the relevant regulations.

The Policies call for the abolition of the approval system for the right to sell cars. The Ministry of Commerce in conjunction with the relevant departments is to formulate new measures for the administration of car brand sales. Under the new system, car sellers will be permitted to develop their car sales business within the scope approved by the State Administration for Industry and Commerce.

Sales service

The Policies require car and motorcycle manufacturers to standardize maintenance services. Manufacturers should publicly announce when they will stop making certain models and ensure that after-sales service and a reliable supply of spare parts will be available for such models for a reasonable time.

Investment

Under the Policies, the examination and approval system for domestic and foreign investment in car manufacturers is to be reformed into a dual system whereby certain projects require approval and others solely need to be filed for the record. The establishment of new Sino-foreign joint ventures in automobile manufacturing requires approval by the SDRC. The Chinese party in a Sino-foreign joint venture to manufacture complete cars, special purpose vehicles, agricultural transport vehicles and motorcycles is required to hold a majority share. If such manufacturer is a company limited by shares, one of the Chinese parties is required to hold a controlling stake that is bigger than the aggregate of the equity held by all foreign investors. Foreign investors in enterprises located inside export processing zones which manufacture cars and car engines for export may be permitted to have a majority share upon approval by the State Council. The Policies permit foreign investors to create two or more joint ventures in China to produce the same categories of vehicles, if they join forces with their existing Chinese partners to associate or merge with other companies in China.

Other issues
The Policies designate seven ports that can be used to import completed cars. From 2005, cars to be sold on the PRC market can no longer be warehoused in the bonded zones of these ports.
The Policies also address various issues aimed at developing a consumer-led automotive market that includes (new and used) car retailing, servicing, financing and related activities.

1.3 The Key Question: the relation between the imports of auto components and the chain of domestic suppliers

Globalization of China's production
Compared with their foreign counterparts, China's auto parts industry entered in the new Century with the following prevailing principal characteristics:

- More labour-intensive in nature
- Rudimentary production techniques
- Inconsistent conformity of products
- Little capital formulation

Prior to the penetration of the overseas market, Chinese firms should exploit their lower capital outlay and labour costs to establish a strong standing in the domestic market. There is a steady increase in the number of firms who have attained the prestigious ISO-9000 and the more stringent QS9000 certifications (Shanghai Hui Zhong automobile Product Co., Ltd. and Shanghai GKN Drive Shaft Co., Ltd.).The achievement implies that Chinese auto parts firms are doubling their efforts to attain the necessary credentials to raise their international competitiveness. Many believe China should first focus to establish itself as a world-class auto parts industry. The establishment of an excellent national auto parts network is one of the effective measures to convince auto MNCs to make China into a main base of operations. Ready access to auto parts components would strengthen their bottom lines, without the need to import costly parts. Streamlining their purchasing and supply chain management functions would raise efficiency in their overall production processes.
This is a critical feature of the Chinese development because in the car sector the key question is that the cost of import of auto parts is a significant pressure on foreign firms to improve the manufacturing quality of their Chinese suppliers. Supply issues and the import of auto parts contribute to significantly elevated costs for Chinese manufacturing.Thus Volkswagen estimated the cost of manufacturing a vehicle in China 18% higher than the cost of manufacturing in Germany in 2003.
For this reason, the goal is to upgrade the chain of domestic supply of auto parts through different actions and investments.

Development of innovative production techniques: a new side of technology and managerial techniques transfer

Large auto firms demand higher expectations from their auto parts suppliers today. MNC Auto parts manufactures have devoted a greater chunk of their budget on R&D on product innovation. Many American and European suppliers have allocated 5% of their budget on product research while their Japanese counterparts spend as much as 6%.With huge pouring in of investments, these foreign MNCs have established world-class R&D facilities, which China's auto parts makers lack the ability to match. Many MNCs manifest their comparative advantage in the development of crucial auto assembly parts while the low-valued added assembly stages will be allocated to the Chinese firms. This is especially common among Japanese firms who often adopt a closed door policy on sharing sophisticated of value-added components even to their partners for fear of eroding their competitive advantage.These intermediate sectors form the nucleus of the auto parts industry.To move up the value chain, Chinese firms should establish working ties with more technologically advanced foreign counterparts. Most importantly, the Chinese should emulate the innovative spirit in their R&D if they were to catch up with the rest of the world.This is already happening in the Shanghai area.

Increasing sophistication of auto parts manufacturing

Advanced aviation and electronic technologies have been widely used in auto parts production today.The culture aspects of safety production, energy conservation, and environmental improvement been received with greater importance by auto manufacturers. Electronic devices are fast becoming a common feature of the modern car - systems for electronic fuel injection, anti-lock brake (ABS), airbags, global positioning (GPS) and central locking. Fingerprint identification technology allows drivers to start, stop, and lock their car.With the rapid strides in technology, the automobile of the future will not be a transportation alone but one of a mobile office and entertainment platform as well.The demand for high tech auto parts will remain buoyant in the future.The Beijing Municipal Bureau of Communications (BMBC) had announced in 2002 that a GPS system will be installed in more than 5000 taxis, in order to reduce unnecessary waiting time and alleviate traffic congestion[2].The new device could also serve to enhance the cabby services to Beijing residents who live in the isolated suburban areas where fewer taxis operate.These residents will be able to hail a taxi easily by calling in the Beijing Municipal Taxi Control Centre through the GPS network.This system is designed with a capacity of 100,000 terminal users, the largest application platform in China.

Roman Fischer, chairman and CEO of DaimlerChrysler China, expects production of Mercedes sedans to reach 25,000 units annually by 2012. Other MNSs have plans to allocate the major bulk of the production facilities due to the com-

[2] Beijing currently has over 68,000 taxis.

parative cost advantages and potentially large market for automobiles. Production of Mercedes sedans and other luxury vehicles in China is likely to raise demand for high value-added, sophisticated auto components.As a medium-term goal, Chinese firms should focus on producing these items.

In order to fulfil its ambition to be a world leader in auto parts production, China is planning to develop an excellent auto parts industry. Its auto components must be inexpensive while meeting international standards. Once domestic auto parts companies reach a certain size and financial standing, China's policies would encourage mergers and acquisitions in the industry.This would increase the size of local companies, making them similar to MNC competitors like Delphi, Bosch,Visteon and Denso. Domestic companies are developing more sophisticated production techniques. For example, in an attempt to reduce their capital costs, Fulin and Fushi Machinery Groups plan to share engines and gearboxes in their new vehicles.

In this decade, China will try to shed its image as a low-tech assembler and to move to production of high value-added auto components - ABS, GPS, and systems for electronic fuel injection, airbags, and central locking. This evolution would be slowed by the reluctance of many overseas JV partners to share advanced production technologies. These fears are not unfounded, because Chinese companies may work with their partner's competitor. Each of China's top four carmakers -First Auto Works (FAW), Shanghai Automotive Industry Corp; (SAIC), Guangzhou Auto and Dongfeng Auto - have ties with at least two global manufacturers.

China is endeavouring to produce interchangeable auto components, e.g., parts meant for a Mercedes luxury car also can be used in a Citroen or Opel. This would facilitate mass production while reducing production costs. Non-pollution auto components were also in the works -solar and electric engines for example. The industry is taking steps to ensure that domestic auto parts meet the rigorous standards of international accreditation bodies like the International Organization for Standardization (ISO). Currently only a handful of the larger companies have attained ISO9000 or QS9000 accreditation - Shanghai Automotive, Wanxiang Group, United Automotive Electronic Systems, Shanghai Ek Chor General Machinery and Fuyao Glass.

Some of the larger companies which have attained both are now considering migrating to the more advanced ISO/TS 16949:2002 system.This is essentially a "process approach" which refers to the application of a system of processes within an organization, together with the identification and interactions of these processes and their management. It is driven by the philosophy that it is no longer good enough to document what you do and to do what is documented but also to verify the effectiveness of the organization's processes in meeting customer and internal requirements.[3]

---

[3] Source: Automotive Industry Action Group (AIAG) ISO/TS 16949:2002 *Implementation Guide*

The domestic auto parts companies have pledged to establish their products into well-known international brands along the lines of Delphi and Bosch. However, such brand name will curtails quality and reliability of its product, which China is clearly lacking. Currently, China has very few established brand names in the international arena. Chinese Fuyao Glass is the best known brand in the auto parts industry, with a substantial market share in China, North America, Europe and Japan.

Beyond attracting Western and Japanese MNCs, the lucrative Chinese automobile industry has aroused investor's interests in ASEAN (Association of Southeast Asian Nations). Possessing the most advanced economy in the region, Singapore is one of the few ASEAN counties that can offer attractive working partnership with China in its bid to produce the highly sophisticated parts. A framework on Joint Study Group on Automotive Electronic Parts & Component Industry Cluster (JSG-AEPCIC) has been implemented. Prominent business leaders have participated in this cooperation. Ng Boon Hoo, the CEO of Sunningdale Precision Industries and Gilbert Ong Peng Koon, Chairman of Armstrong Industrial Corporation were appointed as the Chairman and the Vice-Chairman respectively.A cooperative framework was established between the auto sectors in Singapore and Thailand to jointly manufacture and sell high quality auto parts to China. It has become imperative for China to implement necessary changes and harness new foreign technologies effectively, if it would ant to reach its goal of developing a world-class auto components network.

Without a proper modern auto parts industry, China would fall far behind convincing leading auto MCNs.

Marchisio gives also a more comprehensive assessment of the recent developments of the Chinese OES industry[4] taking as an example the Chinese automotive supplier Huaxiang Company. In 2005 20% of the turnover of the company was to be attributed to the increasing exportations. Furthermore, the Huaxiang Company set up an affiliate in Wolfsburg near the headquarters of the Volkswagen group. In general, it can be observed that in 2005 China has become a net exporter of car components. From the Chinese point of view this is an important result as it shows that there has been important improvement in product quality of recent years. In the past the low quality levels made it necessary to import a major part of the components. As a result more than 70% of the overall value was to be attributed to importations.Today due to the improved quality levels new possible scenarios involving Chinese OESs as well as the big first tier suppliers such as Delphi and Bosch are arising.The Huaxiang Company is, in fact, an example for the changing structure of the sector at global level. The fact that the Chinese supply industry has become a net exporter shows the level of maturity of the Chinese OESs.According to Marchisio the future developments include

[4] Marchisio O., *Siderurgia, cambia tutto: il luogo decisivo ora è la Cina*, in "Liberazione", 06.08.2006, p. 7.

- the Chinese OESs with increasing exportations and an increasing share on foreign markets;
- multinational companies (Delphi, Bosch etc.) that shift production processes to China in order to produce for the Chinese market as well as to use China as an export platform;
- Chinese OESs such as Wanxiang that acquire Western companies in order to obtain technological and managerial know-how.

1.4 Can the Internet increase sales?

The internet enables auto parts suppliers to attend to customer needs over the spectrum of time and space.The tool has proved her worthiness for inventory management based on JIT (Just In Time) principles. At the same time, it can foster closer relationships among automotive parts manufacturers, their retailers and customers. Unfortunately for China, the development of a comprehensive Internet sales system is still jot deplorable, as many domestic firms are still lagging behind their American and overseas counterparts in the areas of e-commerce and on-line purchasing and payment systems.
The auto industry's best-known e-commerce network is Automotive Network Exchange (ANX) in the US DaimlerChrysler, Ford, and General Motors, along with more than 450 of their trading partners, are connected to it. Others are Japan's JNX and ENX in Europe. China is planning a similar secure and efficient e-commerce platform.To stay competitive, and to streamline supply chain and purchasing operations, Chinese companies are expected to apply the Internet and other e-commerce technologies in the near future.

1.5 Social institutional constraints and delocalization limitations

The Chinese labour force

Chinese competition in the car sector can be defined unfair as to the labour issues; it is not only a matter of a widespread social injustice and the absence of fundamental labour rights but the deliberated State manipulation of the labour market as a strategic weapon.
China's growth, in general, is hampered by some social factors.The first is the contradiction between the relative higher degree of free circulation of capitals compared to the low level of freedom of the labour market, the labour market is sill not national but fragmented following geographical delimitations; according to some of the interviewed experts
this is particularly critical for the skilled workers and the professionals.This circumstance is a problem also in view of a new phenomenon: the shortage of skilled workers and the consequent phase of "cannibalisation" among firms of the skilled workforce, offering better working conditions (in China the contract is an individual agreement).
The huge availability of millions of people willing to become, at whatever conditions, an industrial worker and to leave the very bad conditions in the coun-

tryside, is a parallel and simultaneous phenomenon.The latter are also affected by the low level of freedom in the labour market.
The basic education is good but it is not enough to become a productive industrial worker, maybe with the exception of very simple jobs: there is a problem of vocational training, mainly on-the-job but this is not enough to fill the gap between countries with a long tradition of industrial work in mechanical engineering; there are productivity and quality problems to overcome.
The interview partner at DaimlerChrysler confirms that the qualification of the Chinese workforce seems to be quite low and that big efforts have to be made to train people. DaimlerChrysler is doing that, exporting to a certain extent the German qualification system.This operation is considered quite difficult as most workers do not stay with companies for many years. Therefore, the decision to invest in training is not easy to make. DaimlerChrysler sends workers to the OEMs for a limited period of time.The Chinese labour market is strongly regulated by governmental agencies and the foreign OEMs cannot influence that at all.
Among the interviewed experts there are different assessments of this; some are very optimistic because of what they consider a very good level of basic education and because of the subjective attitude of workers that are willing and quick to learn; others are stressing the still big gap between education levels.
Chinese managers are in charge of the Human Resources Management. This is necessary because of the language problem. Often the top HR manager will be European or American, but below that level there are Chinese managers who are carefully educated. In the case of the German OEMs they are sent to Germany for a certain period.There are, however, a lot of obstacles when it comes to implementing European or German HR philosophies.

Working conditions and salaries

In the case of the German OEMs the working conditions are quite good since the plants are standardized all over the world. If it is right that China's competitiveness will no longer be limited to low labour costs but will be based more and more on technological development, quality and managerial improvements then it is likely that these processes will lead also to a segmentation of the labour market and to increasing tensions. It cannot be excluded that these tensions will have an impact on salaries and increase the potential of industrial conflict.

Institutional constraints

There are problems concerning the institutional context. For example, what in general is not known is that there is not one Chinese Labour Law, but there are different laws in different provinces.

Union rights

The other relevant problem is the lack of union rights; formally unions exist but they are not a true independent and highly representative organisation, according to the ILO standards.This means that there is no way to avoid running into

unbearable social and working situation with the consequence of social unrest.There are attempts of workers to make a proactive use of the existing unions.
In the case of the German OEMs there are so far no direct contacts between the German works councils and the Chinese trade unions. The World Works Councils (WWCs) at Volkswagen and DaimlerChrysler had to make a decision how to handle China. In both cases the WWCs do not want to include Chinese labour unionists, firstly, because the Chinese trade unions are not considered free trade unions and, secondly, because in general joint venture partners are never represented in the WWCs. The strategy of the WWCs consists of carefully building the contact with the Chinese trade unions.The impression regarding the Chinese trade unions is mixed. In China labour unionists are more caring for all kinds of social benefits, but not for the real interests of people.The trade unions have a very close relationship with the communist party and they seem to be more interested in guaranteeing the competitiveness of the Chinese plants than in representing the workers' interests.
DaimlerChrysler and Volkswagen have also signed International Framework Agreements (IFAs). So far there have not been cases of application in China.The WWCs, however, are aware that China might be a very serious case.
The implementation of the IFA took place by sending the Chinese version to the HR management and unions in order to communicate the whole issue.With regard to the monitoring the WWCs depend on the IMF and NGOs to come up with cases of violations of principles laid down in the IFA.
The WWCs are trying to cooperate with NGOs. In the case of DaimlerChrysler the WWC organised an expert hearing on the IFA principles, inviting several NGOs like ATTAC or Amnesty International, just asking them to have a look at those principles and also supervising them. So far there have not been any cases that were brought up by them.

Environmental problems

The environmental problems like widespread pollution, are also a hindrance factor for growth. In the case of the car industry this factor has been interacting, in a vicious circle, with the problem of the international standards for emission.The Olympic Games to be held in Beijing are now a powerful reason and strong drive to face the pollution problem and this has interacted positively with the problem of setting emission standards in China which in some years should be comparable with the stricter European ones. This is becoming also a drive for innovation. There are clear signs in China for an attempt to leapfrog from the mass production prevailing propulsion systems to the pioneer systems such as hybrids, electric and fuel cells.

2 Impact on European automobile industry

2.1 European OEMs

In 2005, 54.559.023[5] passenger cars were produced worldwide, of which 17.772.480 (32,57%) in European plants and the new registration of passenger cars in Europe was slightly less than 15 million, it means almost 3 million in surplus that should be sold outside Europe. If the worldwide production of European brands is taken into consideration the challenge for European producers is even bigger; for instance in 2005 the production of passenger cars made by German manufacturers abroad was up to almost 5 million, that is roughly half of the overall German production.

In 2005 the worldwide demand for passenger cars was 53 million vehicles. It means that there is a strong unbalance among production and sales data while the pace of productive capacity building, by each historical brand, is increasing in the attempt to widen the market share and therefore increasing the plant's utilisation rate and rising profit margins. The capacity installed is 80 millions cars per year.There is on average a worldwide utilisation rate of the plants around 75%, which means an overprice for the consumers; the average rate is the outcome of very different rates depending on the producers and this is becoming a competitive factor in terms of costs or profits or both. In the meantime new brands, mainly in China are increasing, mostly in joint ventures with western historical brands, new plants or expanding existing ones.The FDI level in China was 60bn dollars in 2004[6].

According to the Economist:"VW's problems are particularly acute. Its European factories have 20% surplus capacity, compared with 16% for Renault-Nissan and just 3% for BMW"[7].

It can be noted also that some geographical areas are net importers, for instance the US, Germany and Europe. One of the consequences is that competition is focusing on big markets with positive trends such as China, India and Russia; but there are very different situations in these markets because in China there is a growing risk of domestic surplus capacity and fierce competition among the producers. According to KPMG: "Concerns about overinvestment [in China] have been around for some time and it appears that domestic demand may not be able to absorb planned production capacity in the near future. It is believed by some that exports could be a potential but partial solution to the issue."[8]

As it will be argued later this is a key point in assessing the China risk on the mid or mid-long term analysis. A process of consolidation and concentration of the OEMs, and OES, is under way as this picture illustrates(Fig. 3).

---

5 The figures of this first paragraph come from VDA data base.

6 "Financial Times" Asia edition: *China goes global.*

7 "The Economist" on the 16th of February 2006.

8 *KPMG: China Automotive and Components Market 2005*, p. 9

Fig. 3 -

Consolidation in the Automotive Industry

The structural change leads to a continuing consolidation. Until 2015 the number of suppliers will half to 2800 and only 10 of the 12 major OEMs will stay independent.

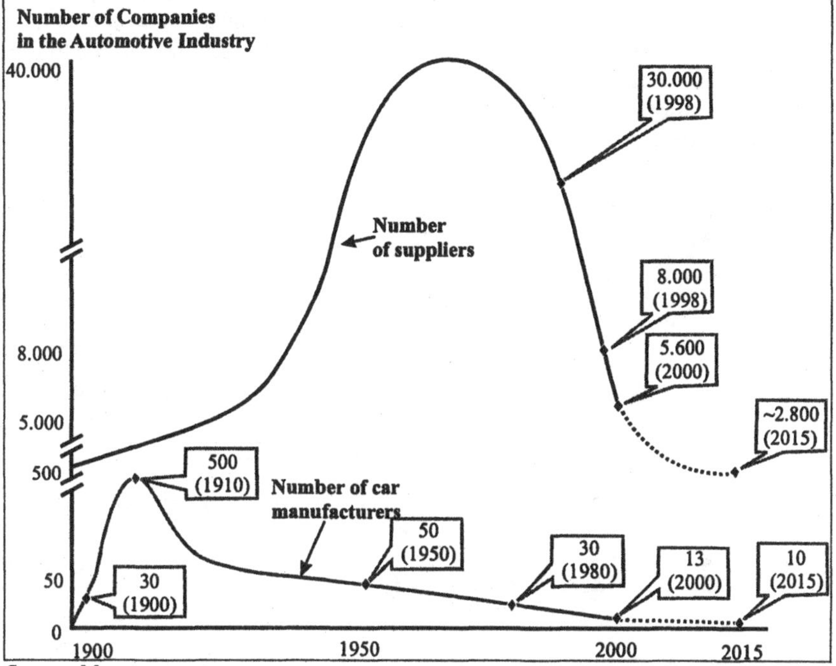

Source: Mercer report

If the agreement that is under discussion between Renault-Nissan and GM will be reached there will be an acceleration of the process. The consolidation process is starting also in China for Chinese OEMs. In any case, out of 54,599,023 cars produced in 2005[9] more than 50% were produced by the first 8 OEMs[10] (Fig. 4).

The competitive scenario for European car producers is different for the different market segments as this picture[11] describes. It is evident that if this forecast

---

[9] VDA data

[10] OICA Statistics - http://www.oica.net/htdocs/Main.htm

[11] from the paper presented by Siegfried Roth, European Metalworker's Federation, Brussels; Chairman of the EMF Automotive Sector Committee at the IPL European conference in Turin (7-8 March 2005): "Relaunching the innovative role of the European automotive industry in the field of technology, organisation and quality of work and The Role of advanced industrial relations Practices".

is correct the European OEMs mostly affected by the international competition in the mid term are:

- OEMs specialised in the volume segment, because of the relative reduction of the market,
- OEMs specialised in the small car/low cost segment because both are growing and the low cost segment will be the first likely target of the newcomers and the small car segment will be the target of almost all the producers, due to the congestion in the metropolitan areas and the likely growing cost of oil.

The case for the premium segment is different because it is expanding also in China (Audi, interview).

Fig. 4 - Radical Change of the Market Structure in the Automotive Industry

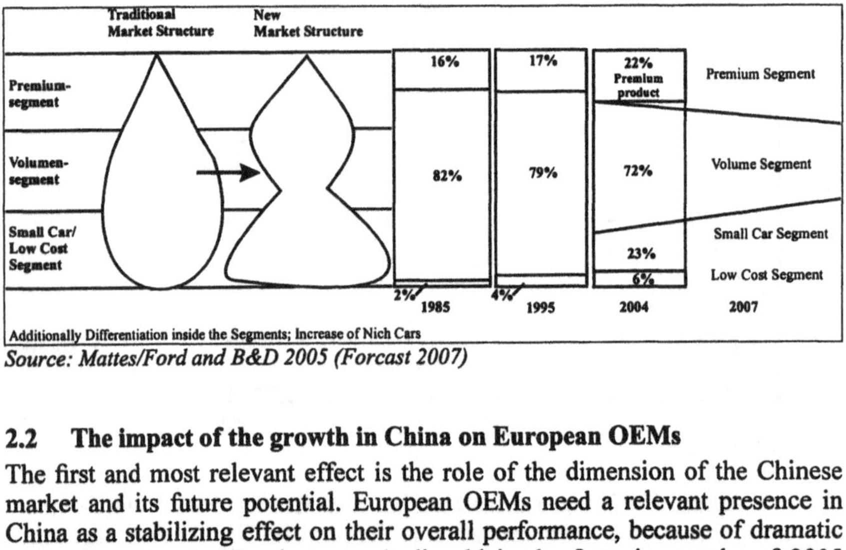

Source: Mattes/Ford and B&D 2005 (Forcast 2007)

2.2 The impact of the growth in China on European OEMs

The first and most relevant effect is the role of the dimension of the Chinese market and its future potential. European OEMs need a relevant presence in China as a stabilizing effect on their overall performance, because of dramatic economies of scale. For instance Audi sold in the first six months of 2005 roughly 12% of its total sale in China. So the market scale will produce, for the time being, different effects on different OEMs, depending on which segments it is focused on and on its market share. All the Western OEMs are convinced that, in the medium run, they will gain acceptable profit margins, also with a low market share, because of the overall market size.

It seems therefore that for European OEMs China is basically a market and not an "export platform".This is confirmed by the interviews with representatives from Volkswagen, DaimlerChrysler and Audi. Only with regard to the Asian and Australian markets China might function as an export platform in the future. In the case of Chrysler cars China might also become a platform for exports to the US market. Given the competitiveness on the US market the management might opt for such a strategy.According to the interviews it is not foreseeable when the first of the US-American car producers will start to import cars from China but it seems to be quite sure that it will happen and every company wants to be prepared for that.

The opposite case is Eastern Europe (within and outside the EU) and Russia.As data clearly show, the main interest is the possibility of utilising these countries as an export platform and secondly to consider their market potential. It is not likely that from 2007 there will be an additional capacity of 1, 2 million cars in these countries. So in terms of possible risks for employment in the EU15 countries it comes from the shift of capacity to the eastern European countries, within and outside the EU. It is clear that the added capacity in the New Member States is not related to the local demand. It is possible to sum up this information in a coherent picture regarding European OEMs. There are several risks for the countries in the former EU15; as The Economist puts it, a push for "the big squeeze"[12] depending on, firstly and mainly, the eastward shift of capacity. China is not a clear and present danger to the EU, as to OEMs, because for the EU-based OEMs China is first of all a market where to sell a part of the overproduction accumulated in Europe (a very small part of it) and where to invest in order to take advantage of the opportunities resulting from the size of the Chinese market.The same can be said for India.

In the medium run China will become, as India, a net worldwide exporter[13] and sooner or later it will try to enter into one of the richest markets of the world, which is Europe.This is, according to the interviews, possible, if nothing will be changed in Europe in the medium – long run. Different is the evaluation of the competitive capacity in other markets such as the Middle East, Maghreb, Africa and also the United States. The interviewed experts think that in these markets the Chinese producers have already become competitive and that their competitiveness will still grow in the future.The reasons for this differentiation are the standards (safety and ecological) and the nature of consumer demand (which priorities among cost, quality, style). In any case China's worldwide ambition will clash with the Japanese presence and "dominance".According to a Global Insight Vehicle Forecast (January 2003) the Japanese cars produced outside Japan, in 2008, will as many as 10–11 millions cars out of a total production of 57 millions cars. If the Japanese cars produced in Japan are added the grand total is 20 million cars.That is 1 out of 3 cars produced worldwide.

The competition will become tough on the market segments that are expanding and it is likely that the worldwide market will be more segmented than now with new differentiations inside the traditional segments. Not only in this industry, a tendency towards "global niches" is appearing, with a new economy of scale in which it is very likely that China and India will play an important role as consumer markets.

On a time scale European OEMs have different milestones to face.

---

[12] Feb 16th 2006.

[13] China is already but for a very tiny amount of cars.

The first is the problem of relocation of plants and employment within the EU. According to the Mercer report of 2003, the EU will see a growing employment level from now to 2015; but the overall figure is composed of a plus on the supplier side and a minus on the OEM side; besides the growth in terms of production volume of passenger vehicles will be of + 73.2 in Eastern Europe and +6,2 in Western Europe.

The second one is the problem of overcapacity; the accepted vision is that this problem can be met by pursuing an aggressive policy on worldwide market shares but what seems very likely is that "worldwide" is a faulty expression. A global market in the meaning of a unified market place is totally unrealistic. There will be global market segments but the supply side of this global demand will be segmented because of the effects of many different factors. The most important ones are:

a. the new division of labour or, with today wording, the new business model of the car industry: "the increasing modularity of products and the great international mobility of firms have brought about a radical reorganisation of production processes at international level: the fragmentation of the value-added chain is reflected in the growing share of parts and components in world trade. This trend, which exploits the modularity of products to capitalise on difference in costs or productivity between the various locations possible, had been on the cards for a long time, but it was the sharply fall in transaction costs (in particular plummeting communication costs) and large-scale opening-up of the South's economies that unlashed all this potential"[14]. Accordingly, the global value chain is articulated in specific segments to maximise the efficiency and efficacy of the mix of costs, quality and logistic opportunities. It means in plane words that it is a unlikely mathematics trying to add and to subtract figures globally; it is possible to have overcapacity in Europe in the midst of plenty in Asia, also for the same OEM. In case this will happen to the same OEM it is possible that losses on one side can be recouped on the other side, in any case it is something that cannot go on for a long time;
b. the necessity of a customisation of products, for functional or cultural reasons (style, symbolic meanings, etc.), to specific geographical areas. The modularity production model can meet this challenge in a very productive way but the segmentation problem will persists;
c. the environmental and legal constraints can be so costly that OEMs can prefer to target with specific products some areas.

So, overcapacity will become a serious problem in a while.

[14] *European industry's place in the International Division of Labour: situation and prospects* – report to Directorate-General for Trade of the European Commission by CEPII-CIREM, July 2004, page 12

The third milestone is the international competition, of which China is a magnifying factor but not the only factor at play. The problem can be expressed in another way as a problem of strategic positioning of Europe in the International Division of Labour.

The only reasonable way for Europe to maintain a strategic role is to change the business model of the European Car industry in a structural way. There are two possible paths to follow, not excluding each other. The first one is already set: to raise the technological content of cars in the premium segment; this strategy will leave part of European OEMs with a competitive margin for a fair time scale.

The interviewed experts are convinced that the task of Europe can only be something that can be done in Europe better than anywhere else. They have in mind top products, top design, luxury goods, machinery, and premium cars. The question is what is it that Europe can offer, what reason do customers have to ask for a European product. Therefore, product innovation is considered so important for the European automotive industry.

The second is the integration of many different industries and economic activities in order to create a new integrated system of sustainable mobility in Europe. It is something depending on the car industry, on urban planning, on new vehicles at the cross road of classical passenger vehicles and light vehicles, on information systems for mobility management, new fuels, new engines, etc. The goal is to create a system for urban areas, that is for the majority of mankind in the near future.

It means that Europe can stay in the international market not only with cars but also with mobility systems, manufacturing included. To design new system of mobility is not only a matter of services or ICT architectures, but also of manufacturing new products in which certain services are built-in.

2.3 The impact of the growth in China on European OESs

As the previous chart has illustrated, the consolidation process is under way also in the OESs field. This process is under way in parallel with a big shift in the value chain of the car industry. The Mercer report forecasts a shift from 65% in 2002 to 77% in 2015.This shift is for Europe very important because according to Mercer there will be, from now to 2015, a regional shift in contribution in the automotive industry with the strongest growth markets in China, India and South-America, that will produce 10 out of the 19 million cars that will be produced until 2015, while Europe achieves the highest absolute growth and becomes the world leading automotive region in terms of added value. But in the NAFTA and Japan the OEM contribution in added value is decreasing strongly. In Europe the decline will be 12 points from 36% to 25%.

On the other hand all the interviewed experts, as well as other reports agree on the fact that China is becoming a "powerful OEM supplier at the global level"[15] in a

[15] ICE (Italian Foreign Trade Institute) and InterChina Consulting – December 2003, page 12.

shorter time and with better chances than to become a global OEM. Is this last statement congruent with the Mercer report? Yes, it is. It is a matter of relative values and pace of growth. In the case of China it would be, in terms of people employed and added value in the OES sector, a jump of roughly 400%, for Europe of roughly 50%. What is still unclear is the Chinese speed in catching up to European engineering sophistication in parts and components.There are different positions concerning this but it should be kept in mind that according to J.D. Powers' research on quality Korea was able to catch up on European and Japanese quality standards in 6 years – from 1998 to 2004 – recovering a difference of 110 scores of 272.According to the interviewed experts the process will not be a jump but a steady and constant process of improvement and of expanding market shares.The logistic constraints for parts and components are not as strict as for cars in terms of costs. Some logistic constraints are due to the actual system of Just-In-Time delivery of parts and components.The new logic of the division of labour

is the new possibilities of disarticulating the value chain in order to distribute single parts of it according to cost/ quality/time opportunities has been fully applied in the supplier side of the car value chain. Accordingly the logistic limitations are very low in the lower segments of the supply value chain and in service related activities[16].

Summing up different options, information and opinions the trend for Europe in the business of supplying OEMs is:

- A growing relevance of OES business in terms of added value and employment
- A shift, through relocation to the New Member States and from the EU to Eastern European countries, also of a part of upper tier parts makers
- With regard to the lower tier parts manufacturing a shift, through relocation and new plants of western parts makers, to China
- A shift to Chinese makers of a part of the supply chain starting from the lower tier parts manufacturing but with an upward trend in the value chain. This trend will be counteracted or limited by some factors.Among these the most important is the necessity for first tier parts makers to be very close to the OEMs plants to make possible the Just-In-Time organisation. For this reason another trend is a faster and bigger shift of services, of EU OEMs, in the supply chain, to China and India.

Thus, the overall effect of these trends and counter - trends is open to two realistic possibilities:

1. a dimension, in terms of added value and employment, of this section of the global car value chain at least proportional to the OEM's activity located in Europe;
2. a diminished and shrinking share.

[16] *China Auto Parts* online: http://en.auto1688.com.cn/newshow.asp?id=4344

In both the bulk of the manufacturing activity will be eastward oriented and a growing part of services will be located in Asia. The latter case will produce a double edge negative effect on European economy as a whole: less added value and less employment.

This trend in the car sector is part of broader trend:

> *Until the mid 1990s European integration saw intra-Community trade accounting fro a growing share of world trade, peaking at 30%. That share since then has fallen considerably and could soon be down to only 20%. This trend contrasts with that observed for NAFTA. There are various explanations for this decline: the emergence of new countries in world production, the incomplete single market, the shift from the integration of production processes at European level to integration at world level.*[17]

It can also be added that:

> *Thus, it is likely that, even at the top end of the range of high technology, new forms of competition will emerge, involving the splitting-up of added value chain. It is perfectly possible for a country of the South if it has a sufficient level of technology, to import components and to assemble upmarket technology products.(..) This development suggests that the reorganisation of the division of labour in technology sectors, with the emerging countries (apart from China and India) being used as platforms for the export of professional capital goods and high-quality components, and China tending to assemble consumer goods.*[18]

According to the interviewed expert Marchisio Chinese manufacturers are already dislocating to Vietnam parts of the value chain.

Marchisio gives also a more comprehensive assessment of the recent developments of the Chinese OES industry[19] taking as an example the Chinese automotive supplier Huaxiang Company. In 2005 20 % of the turnover of the company was to be attributed to the increasing exportations. Furthermore, the Huaxiang Company set up an affiliate inWolfsburg near the headquarters of the Volkswagen group.

In general, it can be observed that in 2005 China has become a net exporter of car components. From the Chinese point of view this is an important result as it shows that there has been an important improvement on the product quality of recent years. In the past the low quality levels made it necessary to import a major part of the components. As a result more than 70 % of the overall value was to be attributed to importations. Today due to the improved quality levels new

---

[17] *European industry's place in the International Division of Labour: situation and prospects* – report to Directorate-General for Trade of the European Commission by CEPII-CIREM, July 2004, page 21

[18] Ibidem, p. 69

[19] Marchisio O., *Siderurgia, cambia tutto: il luogo decisivo ora è la Cina*, in "Liberazione", 06.08.2006, p. 7.

possible scenarios involving Chinese OESs as well as the big first tier suppliers such as Delphi and Bosch are arising. The Huaxiang Company is, in fact, an example for the changing structure of the sector at global level. The fact that the Chinese supply industry has become a net exporter shows the level of maturity of the Chinese OESs. According to Marchisio the future developments regard

a. the Chinese OESs with increasing exportations and an increasing share on foreign markets;
b. multinational companies (Delphi, Bosch etc.) that shift production processes to China in order to produce for the Chinese market as well as to use China as an export platform;
c. Chinese OESs such as Wanxiang that acquire Western companies in order to obtain technological and managerial know-how.

On the basis of these observations Marchisio concludes that China might take on the role of a junction of value chains. The above mentioned developments together with the quantative and qualitative growth of the local market explain the possibility of China taking on in the future the role of an important cross-roads in the supply industry. According to this scenario China's competitive advantage will no longer be limited to low labour costs. The recent developments show that China's competitiveness will be based more and more on technological development, quality and managerial improvements.

These trends might also have an concrete impact on employment levels in the Western European automotive industry in a short to medium-tem period. At present the management of one of the German OEMs is discussing the possibility of outsourcing the internal manufacturing of components. According to a representative of the works council it cannot be excluded that in two or three years the management will try to import components from China. In any case the management will be able to use this potential option to exert a strong pressure on the works council.

From a Western European point of view these overall trends can be managed by institutional policies, for instance by industrial policies choices. It should be noted that according to the CEPII-CIREM report the situation of the automobile sector, still a strong point in the EU, in terms of comparative advantage, is worsening since 1967.

2.4 Recommendations

a. Expand the concept of a fair trade to include the labour issues
b. Utilise the opportunity window of stabilisation and growth for European OEMs deriving from the short-medium term effect of China's growth – and the relative profits – to:
    1. Maintain some strong specialisation such as the premium segment and other possible global niche markets
    2. Move the European sector in the direction of a sustainable mobility: a new generation of engines, new mobility patterns (vehicles and services) for metropolitan areas, developing an ecological dismantling cycle.

3. Utilise the opportunity of growth for European OESs to "harden the core business" (FIEV, Interview), developing the "R" of the "R&D" equation; it means to become top specialists in designing new technical and managerial solutions in the global supply chain of the car sector.

3 Conclusions

1. Expand the concept of fair trade to include the labour issues.
2. Utilise the opportunity of stabilisation and growth for European OEM – and the relative profits:
   2a. Maintain some strong specialisation such as the premium segment
   2b. Move the European sector in the direction of a sustainable mobility: a new generation of engines, new mobility patterns (vehicles and services) for metropolitan areas, developing an ecological dismantling cycle.
3. Utilise the opportunity of growth for European OES to "harden the core business", developing the "R" of the "R&D" equation. It means to become top specialists in designing new solutions for the growing responsibilities of the OES in the global supply chain of the car sector.

Chapter 4
Shipbuilding in China and its impacts on European shipbuilding industry

Thorsten Ludwig, Jochen Tholen

Preface

The present discussion on the development of the Chinese shipbuilding industry and its consequences for world shipbuilding generally and the European shipbuilding in particular is stamped by contradicting appraisals and defective information. One is confronted with different appraisals, however, not only when one questions different actors (e.g., shipyards, banks or suppliers). Also within these groups the appraisals contradict partly considerably. A combination of divergent experiences of the actors, inadequate information and not least existing resentments compared with the " economic power of China " is decisively for it.
This contribution wants to deliver on the basis of expert's interviews as well as the evaluation of appropriate statistics and available literature some concrete information to win a more exact picture of the Chinese shipbuilding industry.

China is second as trade partner with the EU-25; and in 2005 the EU-25 has imported goods from China with a value of 158 billion EUROS . This has direct consequences for the sea-borne trade, and indirect impacts on global shipbuilding.
In 2006 China is the third largest shipbuilder worldwide, and will become Number one in 2015 at the latest.

1 Shipbuilding in China

1. The current stage of Chinese shipbuilding industry
The following graph indicates the quick rise of Far East and – in the last years – especially the Chinese shipbuilding industry. In the 1970s West European shipbuilding participated in the 1970's boom in accordance with its global position at that time. In the period of recession in the 1980s, all shipbuilding regions worldwide suffered by a heavy decrease of orders. In the wake of the two global oil crisis about half of the shipyards worldwide disappeared from the map and thus employment had been halved (Stopford 2004). But in the succeeding recovery in the 1990's and the boom of the global shipbuilding the European yards could not benefit. The real winner were the Far East yards:

Fig. 5 - Deliveries worldwide, regions, in 1,000 gt

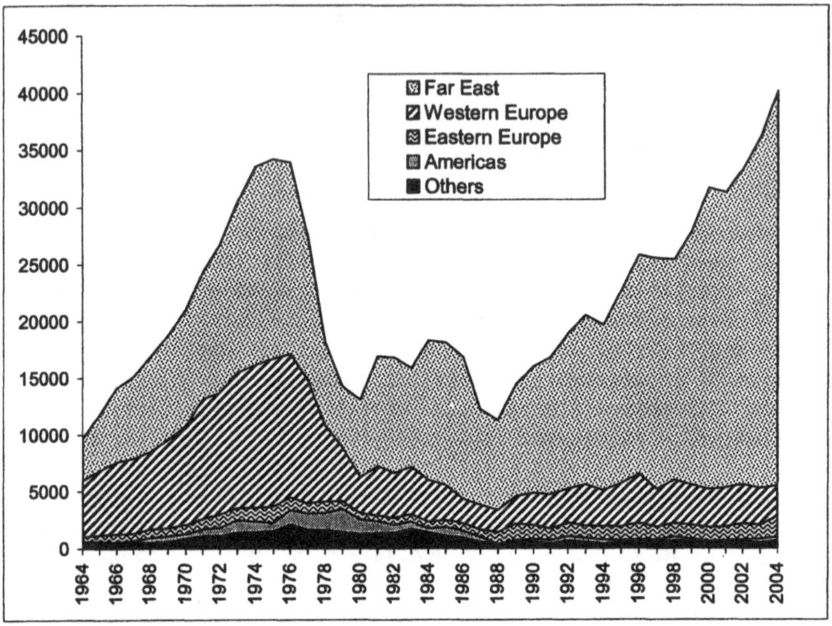

Sources: Lloyd's Register's "World FleetStatistics"

Looking to the direct employment of the global shipbuilding industry, the following figures for China and South Korea did not mirror the exact reality (as f.e. the Korean figures partly comprises the marine supply/equipment industry), but could give an idea of the proportions:

Tab. 11 - Number of direct employment in the shipbuilding industry, by regions, 2004 (Sources: IAW 2006, Suzuki 2006)

China	287,702
South Korea	71,781
Europe (without Russia and Ukraine)	154,872
Japan	40,264

These figures should be treated very carefully, as in the different regions there are different undstandings of employment. Furthermore there are different ratios between direct employment and subcontractors; f.e. in South Korea is ratio is 1:4 , whereas in European shipbuilding industry (with the exception of cruisers) this ratio is 2:1.

The last five years:
China has appeared only in a relatively short time on the map of the world shipbuilding nations. Indeed, the industrial shipbuilding in China has his roots already in the 19-th century (cf. The European Community in 2004, 9pp). At that time the occupying powers began to build up shipbuilding capacities to be able to satisfy their transport needs. Up to this time the Chinese shipbuilding limited itself particularly to the construction of sail and fishing boats. Still today some of the today's most important shipyards in China are located on the locations of these historical shipyards.
The present development of the Chinese shipbuilding occurs after the same pattern as earlier in Japan or Korea. Japan used its shipbuilding industry in the 1950s and 1960s to rebuild the industrial structures of the country. Also Korea looked at the shipbuilding industry as a strategic core for the economic development of the country in the 1970s. China is now in the process of repeating these paths of development, supported by large government investments in the shipbuilding industry (Malhotra 2005).

After the 1950s all shipyards of China had been put under the control of the Ministry of Communication. Since 1982 China State Shipbuilding Corporation (CSSC) is responsible for the Chinese shipbuilding industry. In the 1990s the Chinese government felt constrained to initiate structural reforms in the shipbuilding sector on account of the size of the Chinese shipbuilding industry and the planned entry of China to the World Trade Organisation (WTO). Approximately more than 2,000 shipyards with more than 400,000 employees all together should be controlled no longer through one single conglomerate, because rising inefficiency appears as an increasing problem.

In the 1990s the biggest state transportation companies COSCO and SINOTRANS were also reconstructed. In the context of this reconstruction process many regions and cities also received the control of their own transport infrastructure and smaller shipyards. The most important structural change in the Chinese shipbuilding industry in the past, however, was carried out in 1999. At that time the Chinese government decided to split CSSC into two organizations (CSSC and CSIC) operating independently from each other.
Since then CSSC controls primarily the shipyards to the south of Shanghai, while the newly founded China Shipbuilding Industry corporation (CSIC) is primarily responsible for the shipyards in the north. Partnerships and joint-ventures with foreign enterprises were expressly permitted subsequently to this re-organisation. This, above all, with the purpose to be able to attract the necessary investments and skills from abroad important for the progress of the Chinese shipbuilding industry.

The competitions between the world regions:
In 2005 China was already the third biggest shipbuilding nation worldwide – behind South Korea and Japan. Besides, the pace with which the Chinese shipbuilding industry has entered the group of the world's most important shipbuilding countries is striking. While the Chinese shipyards delivered only 0.9 per cent of all worldwide built ships in 1985, the number has increased to 4.7 per cent only 15 years later. The development between 2000 and 2005 must be emphasized. In this period China has nearly succeeded in tripling its share of the world market. In the same period the share of the world market of the EU-25 halved itself and also the leading shipbuilding nations Korea and Japan had to accept – even if clearly lower – losses with their world market shares:

Tab. 12 - Shares of regions in the world shipbuilding production from 1975 to 2005 in per cent –on GT-base (Source: VSM 2006)

	1975	1980	1985	1990	1995	2000	2004	2005
EU 25	**38.9**	**26.8**	**19.5**	**18.3**	**18.9**	**12.4**	**8.3**	**6.9**
Other European Countries	6.6	9.3	5.7	7.5	4.2	2.4	3.5	2.8
Japan	49.7	46.5	52.3	43.0	41.7	38.2	36.1	35.0
South Korea	1.2	4.0	14.4	21.8	27.8	38.9	36.8	37.7
P.R. of China			**0.9**	**2.3**	**3.3**	**4.7**	**11.6**	**13.8**
Other Countries	3.6	13.6	7.1	7.3	4.1	3.3	3.7	3.9
World	100	100	100	100	100	100	100	100

As in the cases of steel production and the car industry also shipbuilding covers a huge number of different products which differ concerning their technological levels. A comparison between the shipbuilding regions China and Europe has to take this fact into consideration. Since only the total sum of the produced GT still states nothing about the ship types which are produced mainly in the respective regions. Basically it can be found out that the European shipbuilding industry is leading worldwide in the area of highly technological special ships, for example, cruise ships, RoPax-and RoRo ships, mega yachts or high tech naval ships. However, up to now the Chinese shipbuilding industry is concentrated upon the production of standard ships like container ships, oil tankers or bulkers. On the base of the new building orders in the period from 2002 to 2005 (measured in GT) it appears that China and Europe stand in strong competition, above all, in the segment of the container ships. In the viewed period 40.6 per cent of all new building orders in China and 23.1 per cent of all new building orders in Europe were dedicated to container ships. More than 16 per cent of all new building orders in Europe, however, were dedicated to cruise ships and 11.6 per cent to RoRo cargo ships. Until now, in these both segments Chinese shipyards were not able to win appreciable orders.

Tab. 13 - CESA and China - new building orders from the year 2002 to 2005 on GT base (Source:ISL/IAW)

Ship Type	CESA		China	
	2002-2005 in GT	%	2002-2005 in GT	%
Container Vessels	9,254,965	40.6	8,502,083	23.1
Cruise Ships	3,702,191	16.3	0	0.0
RoRo Cargo	2,646,681	11.6	136,915	0.4
Oil Tanker	1,764,585	7.7	8,984,869	24.4
Others	1,683,944	7.4	84,892	0.2
General Cargo Vessel	1,213,951	5.3	1,360,436	3.7
Product Tanker	1,132,273	5.0	4,858,864	13.2
Gas Tanker	669,950	2.9	751,900	2.0
Car Carrier	308,182	1.4	414,000	1.1
Bulker	302,960	1.3	11,538,584	31.3
Chemical Tanker	93,304	0.4	195,809	0.5
Total	**22,772,986**	100.0	**36,828,352**	100.0

The concentration of the European shipbuilding industry on the special shipbuilding (incl. containiership building) is also reflected in the comparison of the generated turnovers of the world shipbuilding nations in 2004. It appears that Europe has produced in 2004 clearly less cgt tonnage than, for example, Japan or South Korea. Anyhow the turnover achieved in Europe thereby was clearly higher than in Japan, South Korea or China.

Fig. 6 - Turnover and tonnage completed of main shipbuilding areas in 2004 (Source: CESA)

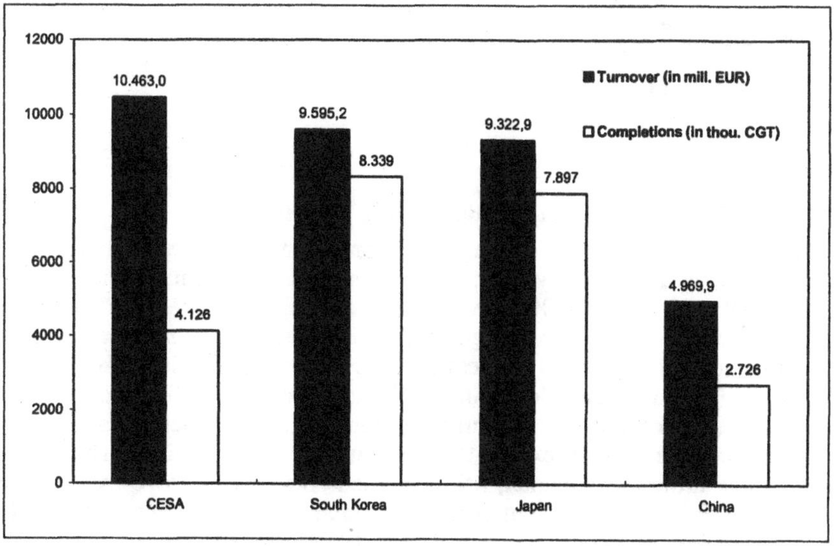

The different levels of turnover in the word regions correspond with the different prices for ship types, that means for example, that the European turnover is 10 per cent higher than the turnover in South Korea (though the completions in CESA countries are roughly half of those in S. Korea), as European shipbuilding industry concentrates of higher value-added ships with higher prices:

Fig. 7 - Prices for merchant ships in July 2005 in Mio US$ (Sources: Coltoncompany, Koshipa, IAW)

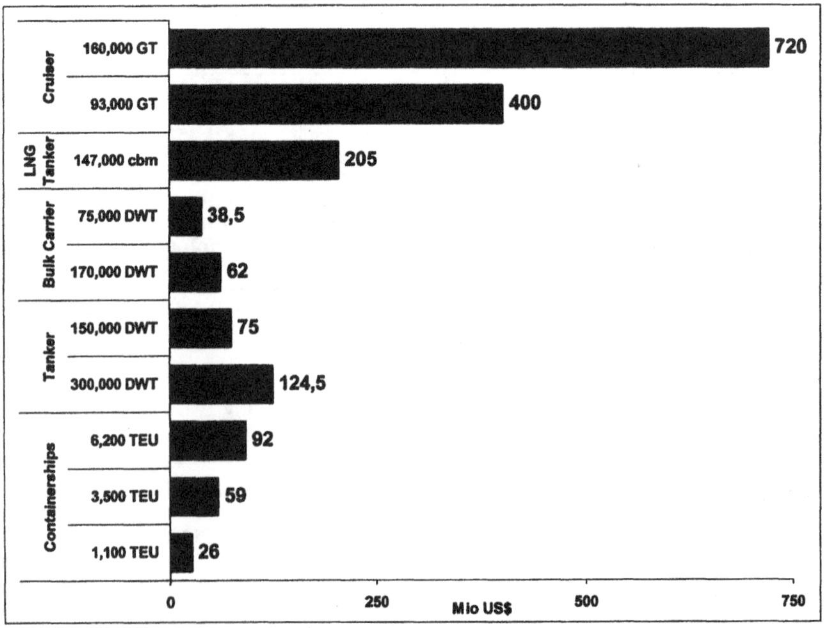

1.2 Structure of the Chinese shipbuilding industry

Until today there exists no dependable information about the number of the Chinese shipyards. Nowadays the number of shipyards in China is estimated at approximately 2,000 – 3,000. Nevertheless, only approximately 600 of these shipyards are able to build ocean going vessels. Classification societies assume from the fact that at most 10 per cent of these shipyards are able to produce for the export (Germanischer Lloyd 2005, 15). As a result of the future consolidation process of the Chinese shipbuilding industry (also by mergers and bankruptcies) western experts expect that the number of the Chinese shipyards, competitive at the world market, will decrease in the next 10-15 years. De Norske Veritas forecasts that in ten years only maximum of 15 Chinese shipyards will exist which compete at the world market (Schmidt 2005, 9).

In spite of the relatively high number of Chinese shipyards the Chinese shipbuilding industry is dominated by the two big groups CSSC and CSIC. Both groups are concentrating more than 60% of the Chinese shipbuilding production with more than 70% of all new building orders placed in China. Due to the outstanding position of these two state owned shipyard groups they should be outlined in the following briefly.

China State Shipbuilding Corporation (CSSC)
CSSC was founded originally in 1982 with the purpose to organise the Chinese shipbuilding sector afore controlled by the Chinese communication ministry. Since the splitting of CSSC in the two groups CSSC and CSIC in 1999, CSSC is responsible for the shipyards located on the east coast of the Yangtze delta and in the southern regions of China.
In 2004 CSSC was already the fourth largest shipbuilding corporation in the world:

Tab. 14 - World Ranking in 2004 (Source: Suzuki 2006, Table 3(2)

Rank	Company name	Construction quantity (in million dwt)
1	Hyundai Heavy Industries	6.93
2	Samsung Heavy Industries	5.36
3	Daewoo Shipbuilding	4.30
4	CSSC	3.57

In 2010 the target is a construction quantity of 10 mio dwt (third in the world) and in 2015 15 mio dwt (first).
All together the group CSSC directly employs more than 95,000 people. The enterprises of the groups CSSC are concentrated in the area around Shanghai, Guangzhou and other cities in the provinces of Jianxi and Anhui. The most important CSSC shipyards are:

- Chengxi Shipyard
- Guangzhou Shipyard International Co. Ltd
- Guangzhou Wenchong
- Hudong Zhonhua Shipbuilding (Group) Co. Ltd
- Jiangnan Shipbuilding (Group) Ltd
- Jiangzhou Shipyard
- Shanghai Shipyard
- Shanghai Waigaoqiao Shipbuilding Co. Ltd
- Tianjin Xingang Shipyard
- Wuhu Xinlian Shipyard Ltd.

A total of 25 big and middle-sized shipyards, 57 maritime supply enterprises, 36 research institutes, three universities and four training centers belong to CSSC. It appears that CSSC is a vertically very diversified corporation. The

shipyards of the group co-operate mostly with maritime supply companies and research institutes belonging to the group.
Since 2003 the China Shipbuilding Trade Co. Ltd. (CSTC) is responsible for the marketing and the sales of ships for the export of CSSC . The single enterprises of the CSSC group have different shareholders but CSSC itself holds, nevertheless, always the biggest share in the single enterprises. The other shareholders are mostly regional institutions or other Chinese enterprises as for example the Chinese steel producer Baosteel.

China Shipbuilding Industry Corporation (CSIC)

After the splitting of the former CSSC in 1999 the responsibility for more than 48 industrial enterprises and 28 research institutes was transferred to the CSIC. CSIC companies are located in more than 20 provinces in the northern part of China, especially in the bay of Bohai. All together approximately 160,000 people are directly employed by CSIC. The most important shipyards of CSIC are:

- Bohai Shipbuilding Heavy Industry Co. Ltd
- Dalian Shipyard Co. Ltd
- Dalian New Shipbuilding Heavy Industries Co. Ltd
- Qingdao Beihai Shipbuilding Heavy Industry Co. Ltd
- Shanhaiguan Shipyard
- Wuchang Shipyard.

As well as in the case of CSSC the marketing and export activities of CSIC were transferred in 2003 to the new founded China Shipbuilding und Offshore International Co. (CSOC). The Dalian New Shipyard belonging to the CSIC group is the first state-controlled Chinese shipyard which was able to win orders for VLCCs (Very Large Crude Carrier). As well as CSSC, CSIC is able to build more than 100 different ship types. This encloses naval as well as merchant ships.
Because the group also owns enterprises which produce not only especially for the maritime sector, CSIC also works in the sectors of airplane construction, metallurgy and energy. Analogously to CSSC, shipyards belonging to the CSIC group, have different shareholders. In most cases the CSIC owns the majority, while the remaining shares are held either by banks, individuals or asset management institutions.

Others

Beside CSSC and CSIC a huge number of other shipyards exist in China. Most of them as for example the Changjiang National Shipping Group or the Fujian Shipbuilding Industry Group Corporation are furthermore under state control, directly or more indirectly. There are still a not exactly verifiable number of shipyards which are in private property. According to estimations of the European Community approximately 10 smaller and middle shipyards were in 2004 in private ownership (European Community in 2004, 12).

The following ranking shows the importance of the single Chinese shipyards, in relation to their orderbooks, specified by the type of ship and the origin of orders (domestic/foreign shipowners) (Fig. 8/Tab. 15)

Tab. 15 - Orderbook of Chinese Shipyards (May 2006)

Shipyard	**Tankers**	**Gas Carriers**	**Off-shore**	**Bulkers**	**Container-ships**	**Other Dry**	**TOTAL**	***% Share***
Dalian Shipbld. Ind.	1,61	-	-	-	1,214	-	2,824	14.0%
Waigaoqiao S/Y	588	-	255	1,019	-	-	1,861	9.2%
Hudong Zhonghua	50	368	-	456	428	119	1,421	7.0%
New Century S/Y	1,221	-	-	-	117	-	1,338	6.6%
Bohai Shipbld.	627	12	-	270	-	-	909	4.5%
Shanghai Chengxi	-	-	-	341	482	-	823	4.1%
Nantong S.B.	286	-	-	236	284	-	806	4.0%
Jiangnan Changxing	491	-	-	-	274	-	765	3.8%
Guangzhou Shpyd	763	-	-	-	-	-	763	3.8%
Jiangsu S/Yard	-	-	-	-	599	75	675	3.3%
Jiangnan S/Yard	161	11	-	460	-	-	632	3.1%
Jinling Shipyd.	327	-	-	-	259	25	611	3.0%
Dayang S.B.	-	59	203	226	79	-	567	2.8%
Zhejiang S.B.	9	-	153	123	226	-	511	2.5%
Qingshan S.Y.	202	-	-	-	182	96	480	2.4%
Guangzhou Wenchong	-	-	8	-	353	-	361	1.8%
Xingang S.Y.	-	-	41	132	-	171	344	1.7%
Zhejiang Yangfan	-	-	-	-	328	-	328	1.6%
Kouan S.B.	-	-	-	46	195	72	312	1.5%
Mawei S.B.	-	-	-	-	250	47	298	1.5%
Xiamen S.B.	-	-	-	-	131	146	277	1.4%
Jiangsu Eastern	-	-	37	-	171	51	259	1.3%
Jiangdu Dadong	157	-	-	35	-	-	192	0.9%
Hantong Shipyard	-	-	-	166	-	-	166	0.8%
Weihai S/yard	7	-	-	-	134	6	147	0.7%
Qingdao Hyundai	83	-	7	-	-	23	113	0.6%
Others	788	83	413	224	264	643	2,414	12.0%
TOTAL CGT	7,372	531	1,116	3,734	5,97	1,476	20,199	100%
No. Vessels	407	21	123	195	351	157	1,254	
DOMESTIC OWNERS								
Total CGT	1,598	380	401	576	1,269	42	4,265	21.1%
No. Vessels	78	7	6	27	42	8	168	13.4%
% Domestic (of cgt)	*21.7%*	*71.4%*	*35.9%*	*15.4%*	*21.3%*	*2.9%*	*21.1%*	
FOREIGN OWNERS								
Total CGT	5,774	152	716	3,157	4,7	1,433	15,933	78.9%
No. Vessels	329	14	117	168	309	149	1,086	86.6%
% Foreign (of cgt)	*78.3%*	*28.6%*	*64.1%*	*84.6%*	*78.7%*	*97.1%*	*78.9%*	

Source: China Intelligence Monthly)

Fig. 8 - The most important shipyards in China

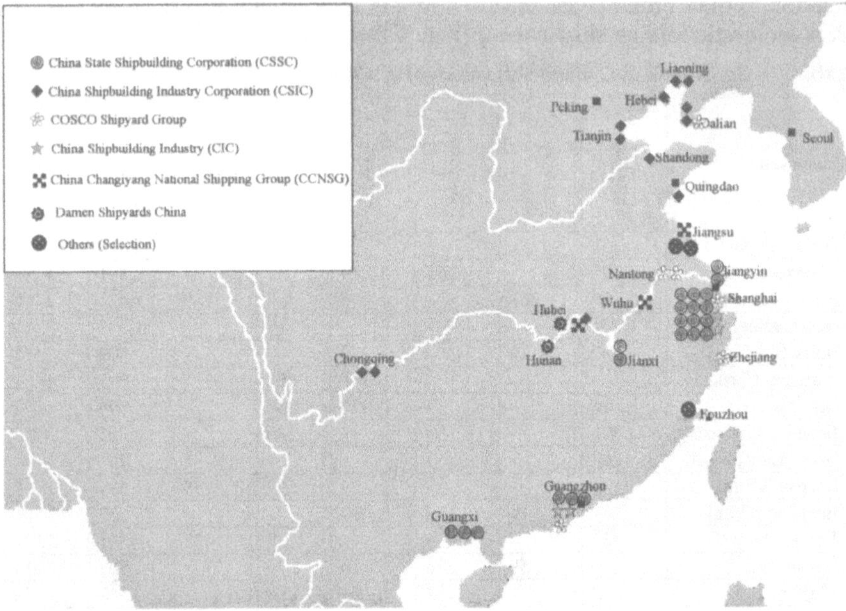

1.3 Forecast

Within the scope of the 5 year plans for the economic development decisive defaults were formulated for the Chinese shipbuilding industry. The State Commission of Science, Technology and Industry for National Defence has set targets for shipbuilding output at 10 million dwt in 2005 and 24 million dwt by 2015, estimated to represent 16 per cent in 2005 and 35 per cent in 2015 respectively of world shipbuilding output. To reach these goals seems to be realistic if one looks the present shipbuilding capacity developments.

The OECD has provided a prediction for the period from 2000 to 2007 concerning the growth rates of the production capacities in world shipbuilding. Only during these seven years China will have raised its shipbuilding capacities by 52 per cent– the biggest increase by far worldwide (Fig. 9)
The South-Korean shipyards increased their production capacities about nearly 18 per cent in this period, while the capacities of the Japanese shipyards will grow about 8 per cent during the same period of time. The shipbuilding capacities in the European CESA member countries increase with 8.5 per cent. Indeed, CESA's groth rate still lies above the value of Japan, nevertheless, South Korea and China stand out with clearly higher growth rates.

Fig. 9 - Proportional change of the world shipbuilding capacities (cgt) from 2000 to 2007

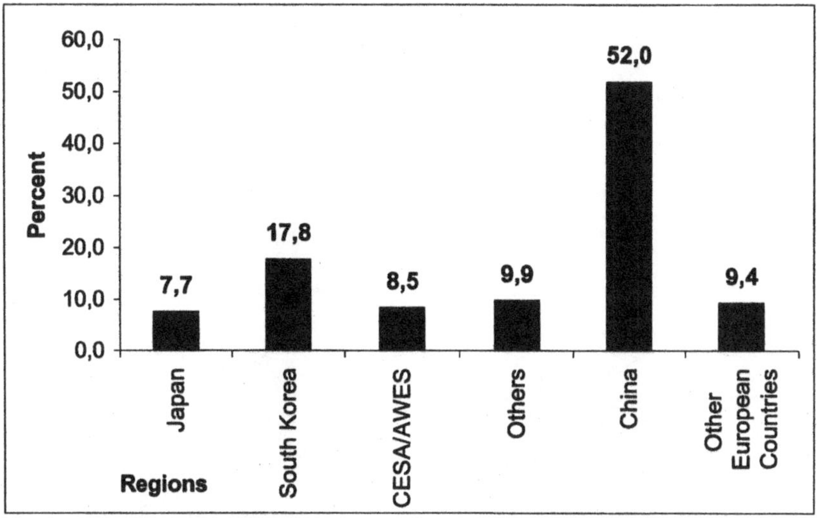

Source: OECD in 2004 b, 16, based on Lloyd's Register Fairplay; own

Indeed, the OECD forecast has examined only the period from 2000 to 2007. Nevertheless, in view of the announced investments in new shipyards it must be assumed that the increase of Chinese shipbuilding capacities will also proceed with the same pace during the next years.

The forecast of the Japanese Shipbuilders' Association indicates that even in the year 2010 China will become Number one in terms of capacities (Fig. 10).

CESA has presented a forecast, which compares the world shipbuilding capacities in 2010 with about 50 mio cgt with the expected demand in 2010 of roughly 24 mio cgt. That means that in 2010 the ration of capacity versus demand will be 2:1. (Fig. 11)

Fig. 10 - Growth of World Shipbuilding Capacities 2004 – 2010

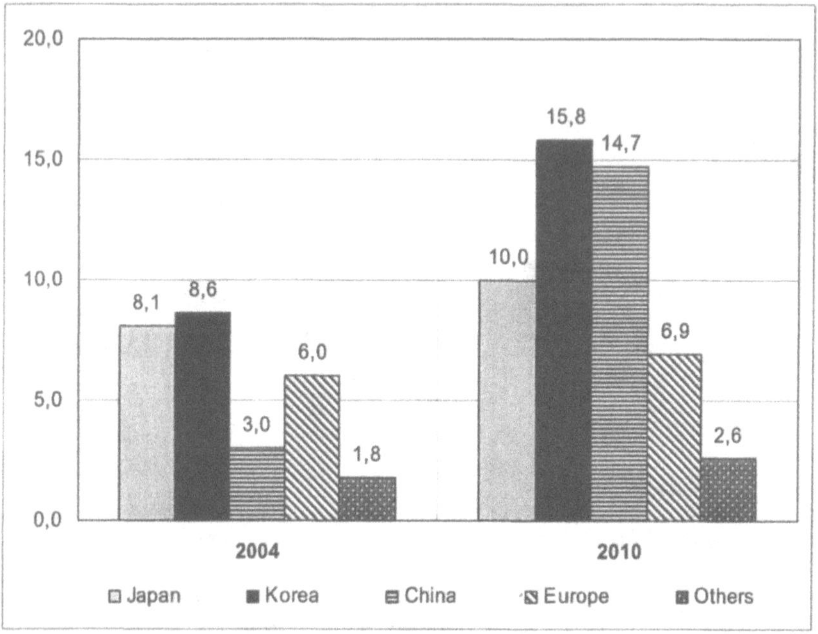

(Source: SAJ)

Fig. 11 - World shipbuilding capacity and demand in 2010

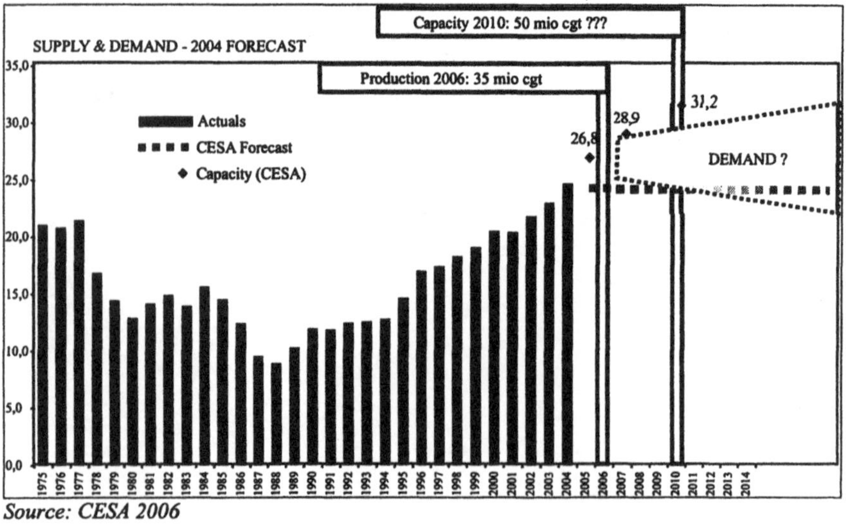

Source: CESA 2006

The existing Chinese shipyards will expand their capacities on locations while new docks and facilities will be constructed in the designated areas. The two state-controlled groups CSSC and CSIC are in the forefront of this expansion. In order to reach the target to become no. 1 of the world's shipbuilders, CSSC is expanding its yards (Tab. 16).

Tab. 16 - State of Facility Expansion at CSSC Member Shipyards (Source: Suzuki 2006 Table 3(3)

Shipyard Name	Facility expansion plan	Completion date	Planned capacity in mllion dwt
Shanghai Waigaoqiao Shipbuilding Co. Ltd	Expansion of capacity in two-phase construction	2008	2.60
Chang Xing Shipyard	Merger of Hudong Shipyard and Jiangnan Shipyard	2015	8
Zhongming Shipyard	Transfer of Shanghai Shipyard	2010	1.5
Nansah Longxue Shipyard	Merger of Guangzhou Wenchong and Guangzhou Shipyard International Co. Ltd	2008	3
			15.10 (Total)

Similar to the expansion of CSSC, the CSIC Member shipyards will expand too and is planning, to be equal to its "taller" brother CSSC:

Tab. 17 - State of Facility Expansion at CSIC Member Shipyards (Source: Suzuki 2006 Table 3(5)

Shipyard name	Facility expansion plan	Completion date	Planned capacity in mio dwt
Dalian New Shipyard	VLCC dock expansion	2009	3.30
Dalian Heavy Industries	New VLCC dock expansion	2010	1.50
Bohai Heavy Industries	New VLCC dock expansion	2007	1.50
Haixi Wan Shipyard	Transfer of Qingdao Beihai Shipyard	2015	4.68
Qinhuangdao Shipyard	Shanghaiguan Shipyard to construct an joint venture shipyard	2010	1.50
Tianjin Binhai Shipyard	Transfer of Xingang Shipyard	2011	3
			15.48 (Total)

CSSC has set its expansion targets very high, aiming to become the 5th largest shipbuilding group in the world by 2005, 3rd largest by 2010 and the largest by 2015. Embedded in this process, it will increase its shipbuilding capacity from

the 3 million dwt tons in 2005 to 15.10 million dwt tons in 2015. In agreement with the Shanghai municipality, 8 kilometres of Changxing Island facing the mouth of the Huangpu river on the Yangtze Delta will be developed to become the largest shipbuilding site in the world at a cost of approximately US$3.5 billion and to be completed around 2010. Funds will be raised mainly through loans and investments in the form of bonds and rights issues. Additional funds will be raised from the sale of land in existing inner-city facilities. The first stage of the development will include the completion of three docks by 2007. This site will take over the orderbook of Jiangnan Shipyard whose current facilities in Shanghai will be offered for development. Part of the facilities in the second stage of development will be used to replace the Hudong Zhonghua facilities, also in Shanghai. By 2015, it is expected that the completed facilities will be able to deliver 8 million dwt tons of ships a year. The facilities, offering seven large building docks, are expected to attract high value orders such as LNG carriers, cruise ships, ferries and offshore structures as well as very large crude oil and bulk carriers.
In the south of China, the CSSC group is planning a new site on Longxue Island near Guangzhou. With an initial investment of US$543.4 million, the yard is expected to be completed by 2008 and will have a shipbuilding capacity of 2.0 million dwt tons annually.
In the north of China, CSIC plans to expand existing facilities in Dalian and Bohai Bay Shipbuilding capacities.
On the marine engine field, special support is given in the production of low speed diesel engines only to existing manufacturers in Dalian, Shanghai and Yichang on the Yangtze River.
Concerning the future development of the Chinese shipbuilding it is above all emphasised that the up to now underdeveloped ancillary structure is the main problem in China. Compared to the main competitors at the world shipbuilding market most analysts agree to the fact that the Chinese shipbuilding industry dates back from approximately five to eight years behind the Korean and the Japanese shipbuilding industry.

Up to now the Chinese shipyards still benefit from the low labour costs, but this advantage could be reduced in the near future: "The steadily rising prices for steel, coal, electricity and transport are dogging the shipbuilding industry. With an average gross profit of less than 10% for shipbuilding in China rising costs will inevitably threaten the profitability of the county's shipbuilding enterprises" (Yundan 2005).
Indeed, today the advantage of low labour costs is already not as big as it is often subordinated. This advantage is partly offset by the number of workers needed to perform tasks already automated in other countries, and of importing steel and parts (Malhotra 2005).
To increase the productivity, some shipyards have gone over to reorganize the working organization as well as the wage structures. Thus in the Dalian New

Shipyard wages depend of the quality of the work. The president of the Dalian New Shipyard (Mr. Sun Bo) said: „It's difficult to control the workers if they get payed whether they work or not" (Brewer 2006). Beside the linking of wage and working quality the workers were grouped in the shipyard also in different departments in teams which compete with each other.
While the most important shipyard groups, CSSC and CSIC, have relative clear developing strategies, analysts assume from that, above all, small and mid-size shipyards will get difficulties in the near future. This is due to the fact, that these shipyards dispose of no own R&D capacities: "For these yards, their only advantage is cheap labour and current strong demand for shipbuilding capacity. In a weaker market environment with strengthened competition, these shipyards will find it hard to survive" (Schmidt 2005, 9).

Provided, the Chinese shipbuilding industry will also develop dynamically during the next years, this will strongly change the structure of Chinese shipbuilding. Furthermore learning curve effects and economies of scale can be expected. The increasing competition within the Chinese shipbuilding industry will force changes in labour productivity. Underlining the fact that many yards are overstaffed, the upgrading of the yards with new technologies and organizational changes will improve productivity, but in some cases significant job losses will occur. For example the Hudong shipyard employees aproximately 17,000 workers (11,000 workers with beige clothes and 6,000 sub-contracters with blue clothes). According to information by the EU Commission almost all of the beige dressed workers are surplus staff, while the main work is done by the subcontractors. Today the yards are not allowed to lay off the surplus workers, but this could change in the future – confronting the government with huge social problems.
Especially the Korean and Japanese´s FDI in Chinese Shipbuilding could lead in the face of overcapacities to two different scenarios: Either the building capacities in the FDI countries of origin could be reduced, which could produce political tensions in the countries concerned. Or the foreign companies would decide to close down their yards in China. In this case the competitive situation of the mother companies would deteriorate, as they now could not exploit the low Chinese labour costs.

1.4 Chinese policy for the shipbuilding sector

The development of the Chinese shipbuilding is often compared to the development of the shipbuilding industry in South Korea. But the conditions for both processes of development differ considerably. The South-Korean shipbuilding industry was aimed from the outset to produce for export. In contrary, the shipbuilding industry in China was developed to strategic reasons as it was the case, for example, in the former Soviet Union. The substantial construction of shipbuilding capacities in China is founded to enable China to manage the sea transport independently and with own transport capacities. Above all, the need for

raw materials for domestic manufacturing, the need for food for the Chinese population and the exports by the Chinese economy should be covered by ships built in China. This is another difference between China and South Korea in this respect: South Korea never needed a domestic fleet of merchant ships for its exports or for importing goods.
Outspoken target of the Chinese administration is to develop China as the leading world shipbuilding nation until 2015 (Tab.18).
To reach this target, the Chinese government has resolved to develop three regional shipbuilding centres in China: the Bohai Bay in the north, the Yangtze Delta in the centre of the coast and Pearl River Mouth in the south. In these three regions the existing shipbuilding capacities should be modernized and new shipyards be built. Locations beyond these main regions should not be promoted in the future. The shipyards in the officially mentioned regions are asked to increase their productivity by improved management processes and raised qualification of the employees. The lowering of the production costs is declared as the main target of these measures.

Tab. 18 Long-Term Plan of the Chinese Government (State Commission of Science, Technology and Industry for National Defense) (units: 10,000 dwt; per cent)

	Construction Quantity	**Share**	**Remarks**
2004	800	14	
2005	1,000	16	2005 actual performance: 12 mio dwt, 18 % share
2010	Shipbuilding power	25	Sharing the markets in Japan, S. Korea and China
2015	2,400	35	China acquires the number one share
2020	Shipbuilding Superpower		China achieves world standards in shipbuilding technology as well and becomes the world leader in both, name and reality.

Facing this ambitious industrial policy the Chinese government has put on a catalogue of industrial-political measures which should support the shipyards:

- The income tax will remit to the shipbuilding enterprises under certain circumstances. If the shipyards have done investments before 2010 in China, they can be exempt from the income tax for maximum five years after the investment. Nevertheless, this tax exemption refers only to the profit which could be generated by the investment. The tax exemption may amount to maximum 5 per cent of the effected investment.

- Export tax rebates: For the construction of ships for the export the Chinese shipyards can assert further discounts with the export tax. Indeed, a list was provided in 2004 for products which are not released any more from the export tax, ships for the export were expressly excluded from this regulation, however.
- Investment funding reforms: Shipbuilding companies are allowed to raise capital for plant and site development from public issues or corporate bond sales. "Other forms of fundraising will be encouraged through ship leasing companies, very similar to the Japanese 'shikumisen' system where trading houses were instrumental in the development of the Japanese shipbuilding industry with finance and charter arrangements. These companies will be controlled by the Chinese authorities, initially attached to large existing corporations already involved in the shipbuilding, transportations or energy sectors (...) but will be encouraged to attract foreign investments. Such companies will order ships for leasing to domestic or foreign operators and will enjoy preferential shipbuilding loan terms and insurance guarantees. Moreover, such leasing companies will enjoy a preferential business tax regime from their start, being exempt from profit tax for the first five years since their establishment and paying only 1 per cent of their net income as tax between years 6 and 10 of their operation" (European Community 2004, 16). The shipyards themselves will be able to finance their activities through government revolving laps, occurred on the work in progress, as assessed by government financial inspectors.
- Stabilisation of material costs: To secure a steady flow of steel supplies for the shipbuilding industry, the government plans to deliver 80 per cent of the required steel by domestic steel producers. In order to reach this target, the government has promised to subsidy the technological modernization of the steel companies.
- Last but not least the government in Beijing aims to be able to deliver at the latest in 2010 80 per cent of all shipbuilding parts by domestic marine suppliers/equipment manufacturers. Therefore, it was allowed to Chinese maritime suppliers to create joint-ventures with foreign partners. In addition, it is also allowed to foreign maritime suppliers to build up own production plants in China which can be to 100 per cent in foreign ownership.
- Foreign investors in new shipbuilding and marine engine manufacturing units are only allowed to own up to 49 per cent of such ventures, with the Chinese partners retaining the majority. Such ventures must also be equipped with shipbuilding technology support by establishing their own R&D units.

In particular after China had entered the WTO in 2001 and in view of these supporting measures the question is whether these instruments agree with the WTO-rules on subsidies. Due to the lack of information this question cannot be answered finally. Nevertheless, the EU Commission doubts are expressed whether all these instruments do agree with the WTO-rules. These doubts seem reasonable, because an unpublished study from 2005 about the financing methods in the Chinese shipbuilding enumerates twelve different financial support measures for Chinese shipyards whose conformity with the WTO rules is questionable in many cases

According to opinion of experts it is absolutely realistic if the Chinese government declares to be the biggest (merchant) shipbuilding nation of the world until 2015.

Firstly shipbuilding industry is very much related with other sectors like steel industry etc. This gives a cluster effect.

Secondly the Chinese economy develops increasingly to an export economy and export mostly runs through sea born trade. China likes to participate in this boom.

The state loans to control the economy (growth rates yearly at 9.8 per cent in 2005 and in the first half-year in 2006 even 10.9%) are paid from the export profits. Additional to this source the profits made by state owned enterprises (SOE), the half-state and private enterprises are re-invested by the enterprises even again – in China there is nowadays a high self-financing rate which is only comparable with the self-financing rate of the West German economy in the 1950s after ending of the Second World War.

China is, at least, the biggest foreign currency owner in the world. Of course there is since 2004 the danger of economic overheating (esp. in the construction and car sector). Consequently the state limited the credits and raised special taxes. But this state policy is thwarted by the high self-financing rate Chinese (privately and state owned) companies which re -invest their high profits unimpressed from the cooling-down policy of the government. And foreign enterprises get anyway no state loans and can invest unhindered.

1.5 Four small case studies

In the following chapter some selected Chinese shipyards should be briefly introduced. Drawing from these "case studies" some general statements on certain points (wages, working relations, trade unions, working security etc.) will be made. The information is descended partly from our interviews in Europe, partly they are taken from secondary sources (IG Metall 2004, Simon Clarke 2003).

Dalian New Heavy Industry Shipbuilding Co.

The company was originally set up by the Russians in 1898, taken over by the Japanese in 1905 and became a state–owned-enterprise (SOE) after liberation. Restructured in the 1980s reforms, and it had 17,000 employees in 1984. In 1990 the old enterprise was divided, the old one retaining 10,000 employees with 4,500 moving to the new one. Under the reforms management got more autonomy and was under pressure to improve quality. This yard became the main export base of the China Shipbuilding Industry Corporation (CSIC), the second largest shipbuilding group after the China State Shipbuilding Corporation (CSSC). In 1994 it did not appear to have confronted over-manning too stringently. More than 20 per cent of employees were on contracts, mostly newcomers since 1984. Payment was based on post, skill, efficiency, years of service plus bonus. The yard is paying good wages.

The new shipyard was established because there was not enough space to build large ships at the old site. It is currently the biggest shipyard in China and part of China Shipbuilding Industry Corporation. It used to be 100 per cent SOE but was restructured in 2001. It builds 100-300,000 tonne vessels. It all looks highly mechanised and automated. There have obviously been huge investments in building and equipping the new shipyard. In August 2001 there 400 engineers working on design & research, it was ISO 9000 certified with two production lines (one body line and one outfitting line).

The old and the new shipyard are independent companies. The old shipyard is still doing all right, so have not had many lay-offs.

The new yard employed in 2002 4,700 directly plus 4,000 - 5,000 through subcontractors. They are served by about 200 small and large supplier enterprises.

They need a lot of labour so they can provide jobs. As an SOE it has strong management, party and trade union systems. The CEO is responsible for production and marketing as the heart of management. The party is in charge for supervision of production and management. The union system represents the interests of labour and the coordination of production and business. Even though they have three systems, they have a common interest: the development of shipbuilding. Although it is an SOE, each worker knows that they have to work hard for income because they know that their income depends on the development of the shipyard, so workers and managers have the same aim. Because they have the same aim, they work together smoothly and the details of the contract between the enterprise and the worker are all done in public, according to particular standards.

Their competitive edge is

- New product development
- Lower labour cost
- Ownership reform helps to promote the performance of the company.

The labour cost is about 15% of the cost of a ship, which is only half of Korea and Japan, so the wage level in the company is a little above the city average.

In the new company, especially after 1996, when production increased from 200,000 ton to 600,000 ton, the labour force was insufficient so they began to sub-contract. It is a commercial contract and therefore wages of their workers are paid by the sub-contractors. No workers were displaced by sub-contractors, it was only to accommodate increases in production.

There are about 2,000 first-line production workers. Their average wage is 1,800-2,000 RMB (180 – 200 Euros). The basic wage is about 1,200 RMB plus a seniority payment of 8 RMB per year of service, plus 200 RMB for skill level, plus overtime, plus piece rate bonus. If they do not make their targets, they lose

their piecework bonuses. Workers are paid by piece rate system. They reformed the wage structure in December 2001, giving more to technicians and key personnel. Because it is an SOE, the wage structure reform was based on models recommended by the government: a post-based wage system.
The Party Secretary and Trade Union Chair are both members of the Supervisory Board.
They belong to the Dalian and the provincial Employers' Confederation, but not to the national CEC (All-Chinese Employers′ Association).
At the corporation level the trade union has six departments and two offices. They do their work under the trade union law. The total trade union membership is 5,997, which includes workers in a couple of subsidiaries. They have a total of 28 full-time officers, 14 at the corporation level, 14 at department levels. Shop and team levels are not full-timers.
Baiscally work of the trade union officers is divided into two categories:

- Protect workers' rights, and motivating workers to work hard in the enterprise
- Help settle labour disputes.

The Chair of Trade Union (TU) has been TU official for 29 years. Previously he was the head of department of women's issues at the old factory. He was originally an ordinary worker.
In an SOE the trade union chair is not directly elected. Candidates are recommended by the party organization and approved by the corporation, to check whether the candidate is qualified, they {presumably the Party} listen to the views of insiders and select one to go to the Workers Representative Congress and by a secret ballot the final selection is made.
The role of the Party is to ensure, guarantee and supervise operation of the enterprise, while the role of TU is to motivate workers to work hard for the enterprise and present workers' opinions to the party and the management. The TU represents the workers in signing and implementing the collective contract. In the company contract (CC) all aspects related to the life and work of workers are included.
The CC is published in the factory newspaper: It is signed by the General Director and the Trade Union leader.
Wage agreements will be "negotiated" every year, whereas CCs every 3 – 5 years. Every year there is a joint review of implementation of CC and they might amend it. They negotiate wages every year.
Unions gather opinions of workers through Workers Representative Committee (WRC) twice a year to discuss major issues. They also have a collective contract supervision group which reports to the trade union and the WRC. They send the drafts to individual workers and comments come back from team leaders.
There are very few disputes between workers and management.
The main problems facing the trade union in the next five years are that, due to ownership reform, the company is now regulated by the Corporate Law and not by Enterprise Law. The Corporate Law has different provisions on the WRC.

Under the Enterprise Law the WRC has a right to determine management decision. Under Corporate Law, the company is obliged just to listen to views of workers through WRC. There is a problem with this and it is being discussed in Beijing and they are going to revise the Corporate Law.
With Chinese entry into WTO they face the challenge of improving workers' professional quality.

Shanghai Waigaoqiao Shipbuilding
The majority of the shipyard's shares are owned by the CSSC group. The yard conceives as a great shipyard according to Korean model, and is located in the Pudong New District close to Shanghai on the Yangtze riverbank opposite to the Waigaoqiao foreign trade zone.
The shipyard was founded within the 9-th five-year plan (1996 – 2000) by the All-Chinese Council of State as a " priority industrial project “. Built in two construction steps, it had in the end of 2004 3,900 people. The shipyard is designated to build all kinds of big and also high tech ships, for example, 300,000 TDW VLCC, 170,000 TDW Capesize Bulkcarrier; 160,000 TDW Suezmax and 100,000 tankers TDW Aframax. The annual capacity amounts to 1.8 Mil. TDW on a surface of 2.1 Mio. square metres in a 1,500-metre-long bank stripe. Till 2010 the annual capacity of 2.6 Mil. TDW should be increased.
The permanent staff is held as low as possible (balancing the production variations). To meet these imbalances several thousand of temporary workers are hired who can be dismissed if needed at any time. A temporary worker earns (in 2004) 70 RMB (approximately 7 euros) per day, so that his monthly income can amount to from 1,200 to 1,500 RMB (120 to 150 euros).
As a state showcase the shipyard worries about social interests and the continuous education/qualification of its employees: It offers cheap flats for renting, free lunch as well as sports possibilities. A training centre with connected libraries is available.

Hudong-Zhonghua Shipbuilding
Likewise member of the group CSSC, this shipyard originated in 2001 from the merger of the in 70 years old shipyards Hudong Shipbuilding Group and Zhonghua Shipyard which are located opposite to each other on the river banks of the Huanpu river in Pudong near Shanghai. This new shipyard disposes of an area of 1.35 Mil. square metres in 2,000 long bank lines and can produce 1.2 Mil. TDW annually.
It builds merchant and naval ships, big steel constructions (e.g., for bridges) and is number one in China for heavy diesel engines. Hudong-Zhonghua Shipbuilding is specified on the construction of big container ships (up to 8,100 TEU).
To the Hudong-Zhonghua group also belong Hudong Heavy Machinery (licences from MAN B&W as well as Sulzer), Edwards Shipbuilding (German Chinese joint-venture), Dong Ding Steel Structure as well as more than 100 producers of ships' chandlers, shipbuilding suppliers as well as sub contractors.

All together the group employs 14,000 people, including 2,000 technical employees/engineers for R&D as well as for design.
In addition to that it must be noted that the 11,000 of the Hudong-Zhonghua Shipbuilding direct employees cannot be dismissed. The real work is done by the 6,000 by sub-contracted workers who are to be distinguished from the rest by their blue overalls.
Subsidized lodgings are made available to the employees of the Hudong-Zhonghua group for 40 RMB per month (approximately 4 euros) in which several persons divide a room. With a five-day week the shipyard is paying monthly between 2,000 and 3,000 RMB (200 – 300 euros).

Jingling Shipyard
JIingling Shipyard belongs to the CCNSG group, the third-biggest shipbuilding group in China after CSSC and CSIC. Is located in Nanjing (inland China)on the shore of the Yangtze river on a 300,000 square metre area – but only on 26,000 square metres ships are built.
Fonded in 1952 as a repair shipyard, the Jingling yard later shifted its main focus later to the construction of inland navigation vessels which were produced shortly after beginning of the Deng Xiaoping economic reforms in the late 1970s (the vessels are exported). Since 1995 the shipyard is also present on the international market for sea going ships (oil tanker, chemical tanker, RoRo ships). However, for the production process the conditions of the Yangtze river show special difficulties: The water level of the river allows only during a restricted period of the year the ships to be launched and to go to the coast.
The ship design is not part of the shipyard: Either it gets it from (also European) customers or it is buying the design from external design engineering companies.
The shipyard has an annual capacity of 100,000 TDW and occupies approximately 2,800 people in the annual average.
The average monthly earning lies with 1,500 RMB (150 euros) with a 6-days week, 7h/day.
There is not a holiday entitlement after the legally guaranteed holiday days for the shipyard employees.
The company trade union to which all employees belong organises competitions and training courses on the subject Industrial Safety.

1.6 Some common problems, drafted from the case studies

Wages
Compared to other manufacturing sectors, the wages in the Chinese shipbuilding industry are relatively low. This surprises as shipyards are located in the coastal provinces of China in which comparatively the highest wages are paid – the average annual pro-head income in 2004 in Shanghai is 2,800 USD, while in the province of Hebei (inland China) only 1,100 USD were earned (German Bank Research 2005, 18). But on account of the millions of migrant workers

(respected number 150 millions, see below) there are no problems to recruit enough manpower for this industry.
Also the minimum wages in China – which are fixed by the national government and define really only least terms – vary from province to province (from 250 to 400 RMB per month = 25 – 40 euros).
Fact is, that according to the World Bank China's poor grew poorer in the last years at a time when the country was growing wealthier: The real income of the poorest 10 per cent of China's 1.3bn people fell by 2.4 per cent in the two years to 2003, the analysis showed, a period when the economy was growing by almost 10 per cent a year (Financial Times, 22 November 2006, 3).
West-European enterprises in China mostly pay a little higher wages and offer better social achievements than in the regional average. Also the terms of employment are better here. They try to limit the high operational fluctuation (see below). Compared with the West Europeans the US-American and Far Eastern investors (from Japan, Korea, Singapore and Taiwan) have the reputation of "bad exploiters" who do not follow regulations.

Labour Relations/Collective Bargaining/Unions
The All-Chinese Confederation of Trade Unions(ACTU) founded in 1925 is the transmission strap between party and the working class according to Lenin' s doctrine. On the other hand, it should represent not only the rights of the employees in state-owned enterprises (SOE) and (since 2001) in private companies and joint ventures, but the interests of migrant workers too – a balancing act which cannot not succeed. Historically the members' base of the ACTU lies in the (shrinking) state enterprises. According to information given by the ACTU itself (which must be doubted) the number of members is 134 million people. According to the industrial trade union responsible for the shipbuilding industry, out of a total labour force of 260,000 people working in the 2 major Chinese shipbuilding groups (CSSC and CSIC), 255,000 employees are trade union members (nearly 100 per cent).
From 2004 onwards the financial support of the trade unions was limited by party and state. A process of the limited "cord cramping" of the trade unions from the party and the de-ideologisation of the mass organisations started.
Although strikes are not unlawful, Chinese workers have no strike right – neither this right is regulated in the wage agreement, nor it is fixed by law.
On the contrary, wildcat strikes are strictly punished by the state authorities – in comparison for example with Vietnam where these forms of industrial dispute are also indeed forbidden, but not pursued.
In China the enterprise-related wage agreements dominate: They are regulated "consensually"; that is the party does the default, and after that the company management and the company trade unions in state owned enterprises (SOE) sign it. In this area the operational chair of the trade union is the works director or at least the Humane Resource manager as well (all in one person). In the op-

erational practice the wage agreements reflect only the law position and signify no higher wages/ improvements of working conditions.
Juridically however, since 2004 there is also the opportunity to conclude single employment contracts beyond collective employment contracts.

Labour Law
Before 2004 (labour) laws were passed simply by the government or even by the party by decree. Today there exists a proper legislation procedure. Within this legal process, the trade unions have an important voice.
Still in 2006 there should be a new law for employment contract according to which those enterprises, making three years continuously profits, have to pay higher wages.
According to the Chinese labour law from 2004 the permanent employment contract has been removed by time-limited contracts with only restricted protection against unlawful dismissals. As a countermove the observance of the practiced legal regulations by the companies should be better controlled. For example, the working hours and the vacation are regulated by law:

- Maximum amount of working hours is 40 hours per week; 2 rest days per week are intended, one of the two is compulsory.
- There is vacation in the legal holidays (one week in the Chinese New Year, three days around 1st May and three days around the national holiday 1 October). In addition, from 6 to 10 days of vacation after one year of seniority is the rule in Shanghai and in Beijing.

Indeed, these legal regulations are not often kept in the operational reality.

Labour force: Core staff versus migrant worker/Fluctuation of the labour force
In the Chinese shipbuilding industry there is – embedded in the processes of removal and concentration - the trend to reduce the permanent staff and to fill the gaps with migrant workers in the case of extra orders. These migrant workers are cheaper and if needed they can be fired quite easily. The approximately 150 Mil. migrant workers in China who are recruited, above all, from the rural areas always move on the edge of the illegality (stay permissions etc.). Thus only every fifth migrant worker owns an employment contract.
Now, however, there are attempts to integrate these migrant workers into the social security system. Since beginning of 2004 they are encompassed, for example, with in the accident insurance.
These migrant workers as a "supplement" of the permanent staff still aggravate the problem of the weak ties also of the permanent workers to “their” companies. If in a region another enterprise offers even slightly higher wages than in the own enterprise is paid, many employees change immediately to this new enterprise. In particular the quick move of the higher qualified employees (around whom also in China a fight has broken out) signified for many enterprises a strong restriction of their respective production process. Chinese enter-

prises, joint ventures and foreign enterprises try to meet these problems by defining in the employment contracts high barriers for moving. These clauses define compensation payments (which orientate themselves for their part by the educational expenditures of the enterprise for the change-willing employee) which must then be paid by the poaching enterprise.
Here the question arises, to what extent within the processes removal and concentration of the Chinese shipbuilding industry the qualification of the employees can be developed to an extent which can meet the high technical and organizational challenges coming from the production of high tech ships.
Nowadays the organizational problem and the additional costs linked with the permanent integration of new and unqualified employees (see the remarks about migrant workers) are also caught in the production process still by the extremely low labour costs and the high "flexibility" of the employees.

Health and Safety at the workplace
Also in the Chinese shipyards there is a high number of working accidents (see the - in comparison with Europe - low standards of health and safety regulations) – even if not as high as in the mining or the chemical industry.
The new working Health and Safety Law came into force in November 2002, and guarantees the following rights for the employees:

- Clarification about all dangers in the workplace
- Equipment with high-quality health and safety equipment
- Own suggestions to the industrial health and safety
- Walking out in the case of life-menacing situations
- Compensation payments in the case of a working accident.

But like so often in China, the legal reality is different than the law. The paradigm change in China (not any more the single enterprise, but the state does now supervise the working conditions and the industrial health and safety) (still) does not function in the reality: Indeed, the companies have already dissolved very quickly after 2002 their own departments for supervising the health and safety conditions; and the state authorities are not yet so far to take over these tasks.

Social Insurance
In the rural areas the responsibility for the social security still lies with the families. According to the communist ideology in the cities the social security should be provided by the state owned enterprises (SOE). But after the economic reforms and the increasing privatisation of the Chinese economy the Chinese Council of State decided in 1986 to form the social security system independently from the SOEs. But the new law exists only on paper. A enforceable right for social security does not yet exist. It will probably last another 30 to 40 years, until there is a functioning social security system in China.

2 Impact on European shipbuilding industry

The European participation in the Chinese shipbuilding yards moves at a low level. In contrast to European shipyards Japanese and South Korean shipyards are presently the strongest at the Chinese market. The biggest joint-venture with a foreign partner in the area of the shipbuilding is the partnership between COSCO and Kawasaki Heavy Industries from Japan. The number of the European shipyards which are represented actively with partnerships in China can be counted in a hand. Among the rest, there are two shipyards belonging to the Dutch Damen Shipyards Group as well as Shanghai Edwards Shipyard (a joint-venture that belongs to the German shipbroker Treuhand Hansa and the Chinese shipyard Hudong-Zhonghua.
The French shipyard Chantiers de l'Atlantique (in Feburary 2006 bought by the Norwegian group Aker) is providing technical expertise for the building of 132,000 to 147,000 cubic metre LNG carriers by CSSC's Hudong Xhongua shipyard. These LNG carriers are the first of this type to be built in China. According to information from the European Commission approximately 50 experts from the French company are continuously working at the Chinese yard.

Here the question arises, why the especially the West European shipyards (in this case only the merchant ship new building, not the naval shipyards and repair shipyards) do not cooperate with Chinese shipyards as they do, for example, with East European shipyards. One part of the answer lies in the geographical position – the distances between China and Europe are too big to transport ship segments by water or street or railway at a reasonable price to the European shipyards (ship segments are not in small pieces as for example cars).
Another part of the answer refers to possible direct investments of European shipyards in China. Apart from a specific ownership structure of European shipyards (partly in the private property and small sizes of company etc. with only low company capital which, make bigger financial engagement almost impossible) direct investments signify always also the removal and construction of additional shipbuilding capacities and the run-off of Know How. In spite of in harmonization with WTO legislation, there are huge difficulties to enforce protection of intellectual property rights also in court. The Chinese judicial system is dominated by many lay justices and few professional judges, who are badly paid. All this leads to a considerable magnitude of corruption. In comparison to Singapore, Thailand, Vietnam and India, China takes a short-cut by the protection of the intellectual property rights (Asian Intelligence Survey, July 2004), while with regard to the corruption China takes "only" the third place – but still with dramatically bad values behind India and Vietnam. Also in the area of bureaucracy, transparency of the government and the company sector as well as the Standard Corporate Governance, China ranks low together with India and Vietnam (ibid.).

Whether the strategy to stop of the transfer of Know How by the European shipbuilding industry is the right way or not, can be discussed. The transfer of the knowledge from Europe to China happens anyhow – by European ship owners, who order their ships also in China, by the internationally classification societies and by the European marine supply/equipment industry (including the independent design offices), which is strongly export-oriented and allows to produce also in China.
European and primarily German ship owners pass their orders to Chinese shipyards, which are then financed by European banks and classified by European classification societies.

Tab. 19 - Home country of orders with the Chinese shipbuilding in 2002 in per cent

Countries	Per cent
China	32
Germany	16
Iran	10
Norway	7
Greece	7
Denmark	6
Italy	6
Belgium	3
Singapore	3
Sweden	2
Other	8

Source: Craig Hughes/DNV

In 2002 roughly half (47 per cent) of all shipbuilding orders came from European.
In view of menacing overcapacities in the world shipbuilding in certain segments from 2010 onwards, most European shipyards rather concentrate upon a special shipbuilding: RoRo ferries, cruise ships, smaller double hull tankers, but also high tech smaller (TEU 4.000-5.000) container ships which are not built thus by Chinese (and nowadays neither by Japanese nor South Korean) shipyards. The Chinese shipbuilding industry with its immense building capacity concentrates not upon on the special shipbuilding dominated by European shipyards, but upon Bulkcarrier, bigger container ships and tankers. By implementing this strategy the Chinese will hit the Japanese and South Korean shipbuilding industry hard. Thus the South Korean shipbuilding industry has already reacted in June 2006 and has explained that it wants to enter into the market for cruise ships and compete with the Europeans (press release June 2006).
Nevertheless the strategy of the European shipyards – with its concentration on special ship types -, needs a constant advancement of the technology, improvement of the organisation and the rise of the qualification of the shipyards' employees.
In parallel, China above all also concentrates upon the expansion of its own merchant fleet to transport the import and export goods by domestically-owned ships. This could absolutely lead to a misalignment of the competitive levels in the maritime area: Due to the strengthening of the Chinese merchant fleet in the future the global competition could shift more to the transport level – i.e. be-

tween the current dominating European and the Chinese shipping companies. That would mean that the logistic branch could be the battlefield of global competition in this respect.
Quite differently from the European shipbuilding industry, the strongly export-oriented European marine supply/equipment industry is growing. In 2005, for example, 65 per cent of the turnover of the German marine supply/equipment industry derived from export. 23.3 per cent of these export orders came from China, 29.7 per cent from the other EU countries and only 9.8 per cent from South Korea. The rest is divided between the remaining Asian countries, North America and others (see: VDMA shipbuilding and Offshore-supply/equipment industry in 2005). On the other hand this does mean, that 15 per cent of the total turnover of the German marine equipment industry has been exported to China.
In addition, several European maritime suppliers have their own production plants in China. Reasons for these FDI are:

- Low production costs;
- Closeness to the market: As a market for suppliers China plays a bigger and bigger role because China wants to be a world market leader in 2015 in shipbuilding;
- Local content regulations: A certain part of the ship (80 per cent until 2010) must come from the domestic (Chinese) production.

The question whether a European marine supply/equipment industry needs the European shipbuilding as a "national" reference industry for its global survival is judged by the experts very differently and controversially.
In the public discussion about the future of the European Shipbuilding industry the different labour costs in Europe and in China plays a big role

Tab. 20 - Labour costs per month/worker in different shipbuilding nations, 2002, USD

Germany/Netherlands	2,400 USD
Japan	1,800 USD
South Korea	1,400 USD
Poland	800 USD
Croatia	750 USD
Singapore	600 USD
The Ukraine	400 USD
Turkey	400 USD
China (Shanghai)	375 USD
Romania	300 USD
China (interior)	200 USD
India	150 USD

Source: Presentation of Damen Shipyards Group on the Dutch shipbuilding day in the Driebergen/Netherlands, Spring 2003

Of course using only the term “labour costs” is an unsufficient tool for comparisons. The real measurement could be done by “labour costs per unit”, which comprises productivity etc. But these figures are not available. Despite of this, by using only the labour costs, it could still give an idea of the importance of the labour costs in the decision-making process of ship owners to order ships from yards in the different world regions/countries.

Two scenarios for the European shipbuilding/maritime industry:
There are two consequences of the increasing Chinese shipbuilding capacities on the European shipbuilding and maritime industry:

- Increase of the competition situation:
  Worldwide (in particular in China, but also in South Korea and in Vietnam) increasing production capacities in shipbuilding lead to a reinforced competition of the shipyards on orders. The competitive pressure is not only strengthened by the rapid expansion of the shipbuilding capacities in particular in Chinese shipyards, but in addition, by the extremely low labour costs in China and the unclear financing methods (state loans etc.). In expectation of a possible stagnation of the global shipbuilding market it may be supposed that the Chinese shipyards will compete more with Japanese, South Korean and also European shipyards in a special segment. This possible development is based on the acceptance that Chinese shipyards will enter especially the market for technologically simple ship types (like container ships) which are built at the moment especially by Japanese, South Korean and partly also by European yards. Japanese as well as South Korean shipyards could through this development becoming active in segments of technically more sophisticated ship types. This process would create a tougher competition with European shipyards. Thus the South Korean shipbuilding federation explained in June 2006 the entrance of the Korean shipbuilding in the cruise market for the future. But even if the South Korean and Japanese shipbuilding industry would enter the cruiser market, it would not compensate possible losses of market shares to the Chinese shipbuilding industry, as the global market for cruisers is very limited, compared with the global markets for container ships, bulkers, oil tankers etc.
- Defusing of the competition situation:
  China is in 2006 already the fifth largest commercial nation worldwide and will climb in 2020 to number 2 according to predictions of the World Bank. Besides, merchant vessels show the most important means of transportation to export the Chinese goods, to import the raw materials and to organise Internal-Chinese trade (along the high industrialized coastal regions and the Jangtze river). To modernise the fleet, and to create at the same time new transport capacities, China will cover the rising needs for ships from domestic ship production. Therefore, during the next 10 years, highly complicated ships as for example cruise ships are of low interest for the Chinese shipbuilding industry.

Not the field of low tech ship production but the logistic branch could be in the future the "battlefield" of competition between Europe and China.

The Know How transfer as a key element to develop Chinese shipbuilding industry:
Inevitably for the further development of the Chinese shipbuilding is the Know How transfer from Europe to China, by the European shipbuilding industry, the ship owners and banks which build and finance ships in China, by the classification societies, but also by "blueprint trade" of the European shipyards and/or the design offices commissioned.

There are two contradictory trends:
- For the European shipbuilding industry the uncontrolled run-off of Know How is very dangerous,
- For the European marine supply/equipment industry in the medium term this transfer remains without big negative consequences, rather on the contrary: It strengthens at first their global competitive situation.

3 Policy recommendations

What can be the strategy of the European shipbuilding industry facing the Chinese competition?
First to put the main points briefly:
a. Product mix: Intensifying the concentration on the high tech ship segment as well as on other ship types like container ships, double hull tankers etc.
b. Improvements of the vertical relations of the European shipbuilding industry (supply chain)
c. Improvements of the co-operations between shipyards comprising smaller yards as well (horizontal cooperation)
d. Reinforced cooperation in sensitive fields like R&D and qualification of the workforce
e. Development of Pre- and After-Sales Services
f. Leadership 2015 and EU Social Dialogue
g. Consistent EU industrial policy for maritime industry.
In detail there are following recommendations:

- *Mix of production: Intensifying the concentration on the segment of high sophisticated ships (including container ships, double hull tanker etc.):*
  Cruise ships, RoRo vessels, special ships and naval ships are mainly for the export market. Parallel to this European shipbuilding industry has to keep on track to build (high tech) container ships, double hull tankers etc as well. That does mean that the European shipbuilding industry has continuously to improve its competitiveness: Investments in research and development, improvement of the working organization and product organization, constant increase of the qualification of the employees must move the European shipbuilding into the position to stay 2 to 3 years ahead of the Asian competitors.

- *The improvement of the vertical relations of the European shipbuilding industry with the marine supply/equipment industry:*
  On average 1/3 of the added value of a ship is produced by the shipyard itself, while 2/3 are produced by the suppliers. Formerly integrated shipyards, which covered the whole production process of a ship, are no longer the standard model. Today shipyards are flexible enterprises which are tied together in value added chains with external suppliers. The focus is the creation of the interfaces between yards, their service partners and marine suppliers/equipment manufacturers. Nowadays many suppliers are not sufficiently recognized by the European shipyards as a partner, but as compliable contractors to whom the dependence on the shipyards is made clear again and again. This must change very quickly.
  But an optimization of the Supply-Chain-Management is not sufficient to realise the requirements of the market on appointment (loyalty, quality and costs).

- *Horizontal cooperation between shipyards must be improved:*
  Lacking cooperation between the shipyards (horizontal cooperation) generate structural deficits of European shipyards in the global competition, caused by the split and extremely different ownership structures within the European shipbuilding industry.
  Shipbuilding is a high technology and a system industry, where the system ability and the interlinking of the industry must stand in the foreground. Between the enterprises of the maritime industry, different relations and dependencies exist. Networks, branches, enterprise groups are an important basis of its dynamic. Efficient supplier's-buyer's connections are a determining location factor. Strategic alliances with universities, research institutions, developing institutions and consultation institutions also contribute to the competitiveness. Those shipyards in Europe (and this is the majority) which do not cooperate with each other, will get into considerable difficulties after the prospected lack of the new building boom especially concerning container ships (headwords: global over-capacities, global predatory competitions).
  However, there are also shipyards which co-operate with each other (mostly with yards which belong to the same group/owner) and are very successful in the global competition (e.g., Aker, Damen, TKMS). The European-wide cooperation which is not organized within a group has mostly the character of the relocation of labour-intensive production into low wage countries especially in Eastern Europe. This deals less with cooperation as a win-win situation rather with an outsourcing of work into cheap labour cost countries and expresses a split within the European shipbuilding industry between West and East.

- *A reinforced cooperation also in "sensitive" areas (R&D, qualification etc.) is necessary:*
There are already examples for cooperation in these fields like the "Euroyards"-platform[1]. Cooperation functions in these areas, above all, on the basis of trust cultures, which have their roots in personal contacts. These trust cultures across the borders of the single shipyard should be developed in the future increasingly by the shipyards themselves. Industry federations and also trade unions can play here an accompanying role, allow contacts and create rooms in which mutual trust can originate.
- *Development of a Pre-and-after-Sales service by the European shipbuilding*:
Not only the purchase of (technologically sophisticated) ships should stand alone on agenda of the yards, but financing, harbour infrastructure and worldwide repair services should be integrated into a complete package and be offered to the customers. On this basis the customer can value the whole logistics service.
- *A promising start is the initiative "LeaderShip 2015" of the European shipbuilding industry*
The initiative from 2003 defined in a booklet[2] 30 concrete recommendations in areas ranging from trade policies to industrial consolidation are made. Translating these recommendations into action is helping to ensure a prosperous future for European shipyards and marine equipment manufacturers. The work undertaken represents one of the first applications of the Commission's renewed approach to industrial policy in a specific manufacturing sector. Leadership 2015 has been adopted by the EU Commission.
But Leadership 2015 should be reversed in a way, that practical solutions, especially on the national and regional levels, should be implemented.
However the EU as such has no consistent industrial policy for the shipbuilding industry. Each member state is organizing the policy on its own. That is why the creation of a *"European Maritime Coordinator"* is necessary. The coordinator's tasks could be coordinating all the EU financed programmes related to maritime industry, partly individualized in the various DGs.
- *Another promising approach is the EU Social Dialogue Shipbuilding:*
Both employers and trade unions in the member states as well as on European level, share a common view on the perspectives and challenges that the European shipyards are facing. In order to strengthen the dialogue and implementa-

---

1 The aim of EUROYARDS is to foster the interests of its partners, both by adopting common positions before international bodies promulgate the policies of the sector, and by implementing through working teams created ad hoc, projects in the field of Research &Development and purchasing policy, in order to cut costs. EUROYARDS represents the following major shipbuilding companies in five European countries: Aker Yards Oy (Finland), Aker Yards S.A. (France), Fincantieri S,p.A. (Italy), HDW AG (Germany), Navantia S.L. (Spain) and Meyer Werft GmbH (Germany).

2 LeaderSHIP 2015 – *Defining The Future of the European Shipbuilding and Ship Repair Industry*

tion Leadership 2015 recommendations, the European Social Dialogue Committee was formed in September 2003 in which CESA represents the employers′ side and the EMF the employees′/trade unions. Supported by the European Commission DG Employment and Social Affairs, the Committee concentrated its work in the expiring year on different topics from scientific surveys and high-profile conferences at EU level to widespread events at national level sparking the interest young graduates and highly skilled workers in working for the shipbuilding and ship repair industry. There are four working groups performed the following activities:

- Shipbuilding Survey: In 2004-2005 the University of Bremen/Institute Labour and Economy-IAW carried out a survey of the European shipyards. The document is available in seven European languages and a database for further policies. A panel study is planned in 2008.
- Training und Qualification forum: In October 2005 a seminar took place in Trieste/Italy to get information about the different schemes of qualification and training in European shipbuilding countries/industry. Qualification is one of the key tools to meet the global competition successfully.
- Shipyard Week: The Committee initiated the first European Shipyard Week from 27 to 31 March 2006 featuring numerous evens to promote the sector all over Europe. The week started with a well attended Social Partner Conference at the European Parliament on 22 March in which distinguished persons from the European Parliament as well as from the European Commission, from the employers′ and from trade unions′ side as well as well-known academics participated.
- "Tool Box" Project: The idea was to work on an inventory of tools on how to cope with the ups and downs of industry demand cycles while maintaining skills and employment. This project was carried out by the European Monitoring Centre for Change (EMCC) at the Dublin based Foundation for the Improvement of Working and Living Conditions.

Chapter 5
Steel

Daniela Gradinaru, Philippe Morvannou

1 The Chinese steel industry: How strong, How much excess, How much sustainable?

1.1 Steel, an element of the Chinese economic growth

From the Chinese authorities' point of view, the 11th five-year plan starting in 2006 is the key step for China to become a powerful actor in the steel manufacturing industry.The announced slowdown of the Chinese economic growth did not occur in 2005, with a growth rate of more than 10.1%. In 2006, the growth is still sustained and close to a two-figure rate.

The Chinese economic growth, which was mainly based on infrastructure, real-estate, industrial investments and exportations has been strengthened in 2005 thanks to an increase of the domestic consumption of 12%.

For the first time, the surplus of the trade balance went above 100 billion US dollars in 2005 i.e. a 69.9 billion US dollars increase in one year.

Over the past three years, China has become the world's first metal consumer.

Three developments will influence the future demand for steel:

- Economic growth will always depend more on domestic demand, investment and exports will not be the driving forces behind the economy, which leads to the need for a coordination between domestic and foreign demands.
- Economic growth will also depend on the rate of restructuring and of industry modernization.
- Economic growth must become more and more efficient to spare raw materials and energy.

The steel production

According to statistics, China produced 352 million tons of steel in 2005, which is an increase of nearly 69.5 million tons (+24.6%) within one year, which is the biggest increase ever noted.

Tab. 21 - Iron and steel production in China

MT	2004	2005	P2006	2006/2005	2005/2004	
Crude steel	282.9	352.4	392	39.6	69.5	24.6%
Pig iron	257.8	330.4	370	39.6	72.7	28.2%
Steel product	319.8	396.9	440	43.1	77.2	24.1%
Coke	187.1	232.8			45.7	24.4%
Iron ore	335.5	420.5			85.0	25.3%
Iron a loy	9.2	10.7			1.5	16.0%

Source: Metal Bulletin's 4th Far East Steel Conference

In the meantime, pig-iron production has reached 330.4 million tons, which is equivalent to 28% more than in 2004.
The growth in raw material production (iron ore and coke) developed at the same rate, +24-25% in one year. In 2006 another increase is estimated, which would bring the Chinese crude steel production up to around 400 million tons, which corresponds to a slowdown of nearly 40% compared with the production for 2005. The demand for steel in 2010 is of 480 million tons. It looks like after an increase of growth in the Chinese steel production starting in 2003, the year 2006 was the beginning of a slowdown.The explanation to the increase of production was mainly the domestic demand and secondly the gradual substitution for imports. In 2005 China produced 52.2% long products (an enormous increase could be observed in 2004, sustained by the buoyant demand in the building sector) and 47.8% of flat products.These ratios are inversed in the European Union which produced 37.6% of long products and 62.4% of flat products (Fig. 12; Fig. 13).

The steel consumption
Official construction work and the industry represent more than 92% of the steel consumption in China. The industry includes the following sectors (Fig. 14):
• Automobile
• Shipbuilding
• Transport and telecommunications
• Energy, water and electric energy
• Infrastructures and real estate

Fig. 12 - Evolution of Chinese production by type of product (MT)

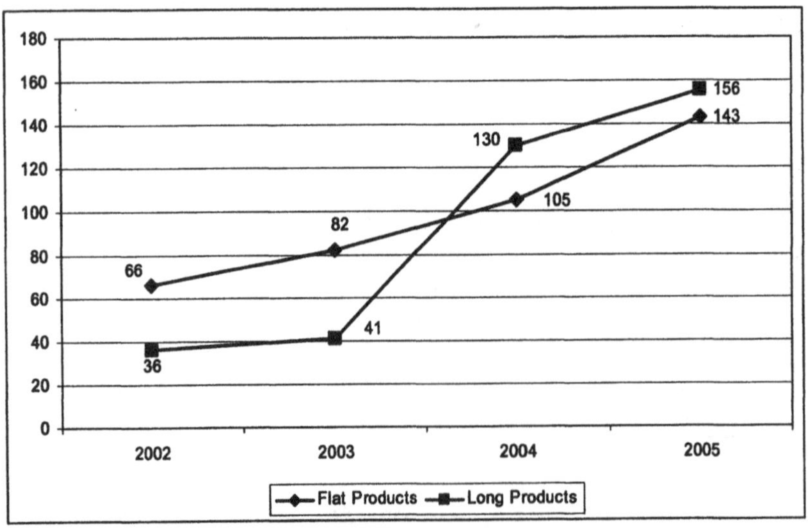

Source: CISA, Baosteel

Fig. 13 - Evolution of EU production by type of product (MT)

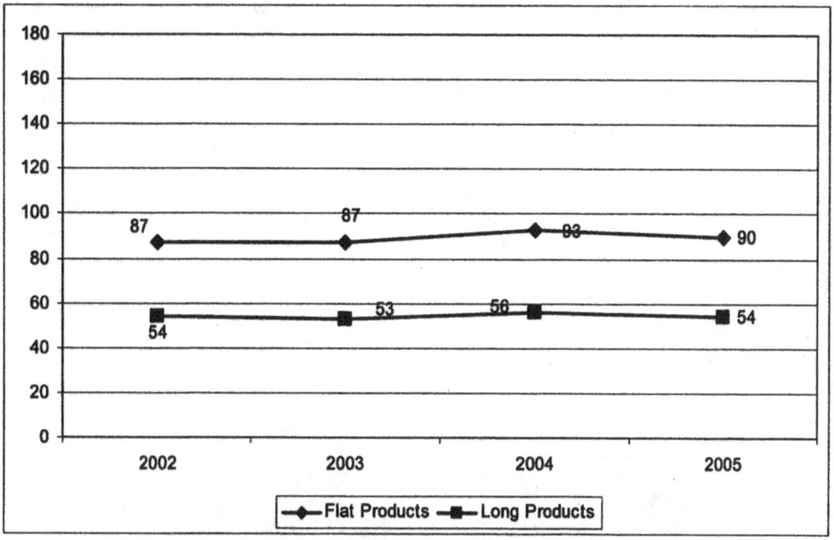

Source: MEPS

Fig. 14 - China's steel major consumers in 2006

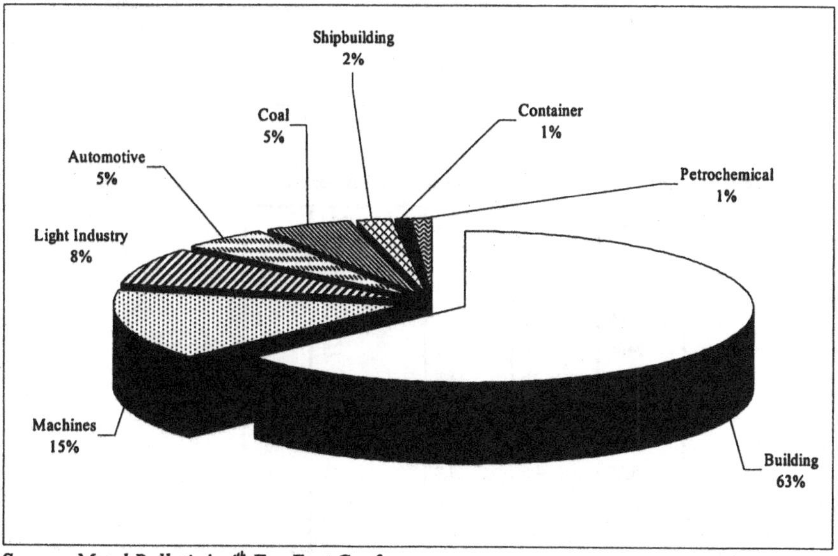

Source: Metal Bulletin's 4th Far East Conference

The growth in these sectors is mainly based on the official construction (63%). In the European Union, this sector uses only 24% of steel production. The automotive sector is the second largest steel consumer in the European Union, with 18% of the consumption. In China the automobile sector is the third largest steel consumer with 5% of the total steel production. For the shipbuilding sector the ratios are almost identical.

Fig. 15 - EU 25 steel consumers QI 2006

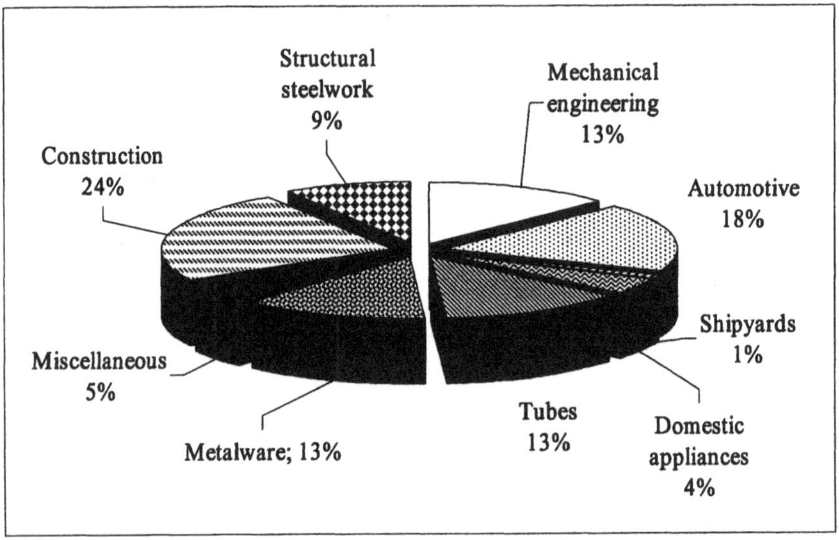

Source: EUROFER

Tab. 22 - Forecast of demand by China's major steel consuming industries in 2006

MT	2004	2005	P2006	2006/2005	
				MT	%
Building	175	204	233	29	14,2
Machines	45	52	57	5	9,6
Light industry	2408	26,6	28,7	2,1	7,9
Automotive	12,76	15,79	18,5	2,71	17,2
Coal	15	16,2	17,4	1,2	7,4
Shipbuilding	3,59	4,4	6	1,6	36,4
Container	3,56	4,5	4,8	0,3	6,7
Petrochemical	2,53	3,33	3,8	0,47	14,1
Total 8 industries	**282,24**	**326,82**	**369,2**	**42,38**	**13,0**
Total China	**296,58**	**352,17**	**400,75**	**48,58**	**13,8**

Source: Chinese presentation at Metal Bullletin 4th Far East Conference

Tab. 23 - Relation between steel and investment in China

Years	Investment in fixed assets		Crude steel (MT)			Steel/invt
	Mds. Yuan	n+i/n%	Consumptions	Production	Net import	T/M Yuan
2001	3721.4	13.0	169.8	51.0	18.7	45.62
2002	4350.0	16.9	205.7	182.3	23.5	47.29
2003	5556.7	27.7	258.6	222.3	36.2	46.53
2004	7047.7	26.6	296.6	282.9	13.7	42.08
2005	8860.4	25.7	352.2	352.4	-0.2	39.75
P2006	10632.5	20.0	400.8	402.8	-2.0	37.69

Source: Chinese presentation at Metal Bulletin 46th Far East Conference; own calculations

The increase in investments over the past years is directly linked to the increase of consumption and of steel production. But the link between the investments in fixed assets and the consumption of steel tends to weaken within this period as indicated in these data. 47 tons of steel were necessary per million Yuan of investments in 2002, 10 tons less are needed in 2006.

Fig. 16 - Steel investment ratio TIMYuan

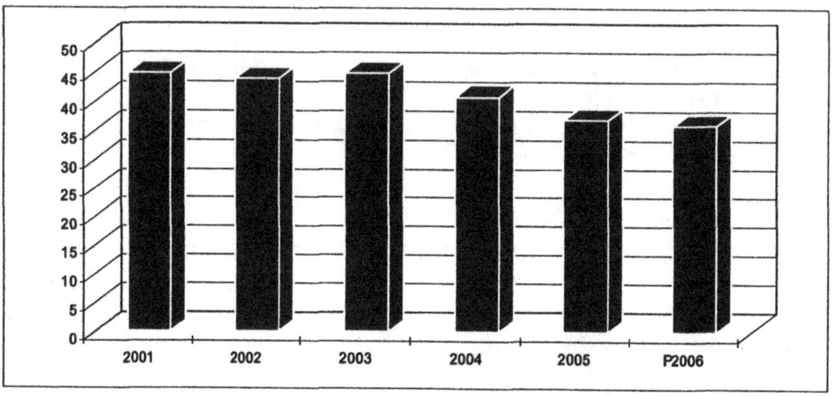

At the same time, the increase in the Chinese domestic consumption slows down the fall of the global ratio between steel consumption and Chinese economic growth. Thus in 2002 the Chinese steel consumption was 20.42 tons per million Yuan of national crude product against 22.06 tons announced for 2006 and 22.57 tons in 2005.

Importations and exports
The balance between production and consumption has been very dynamic in the past years.

Fig. 17 - Consumption, production and import of steel in China

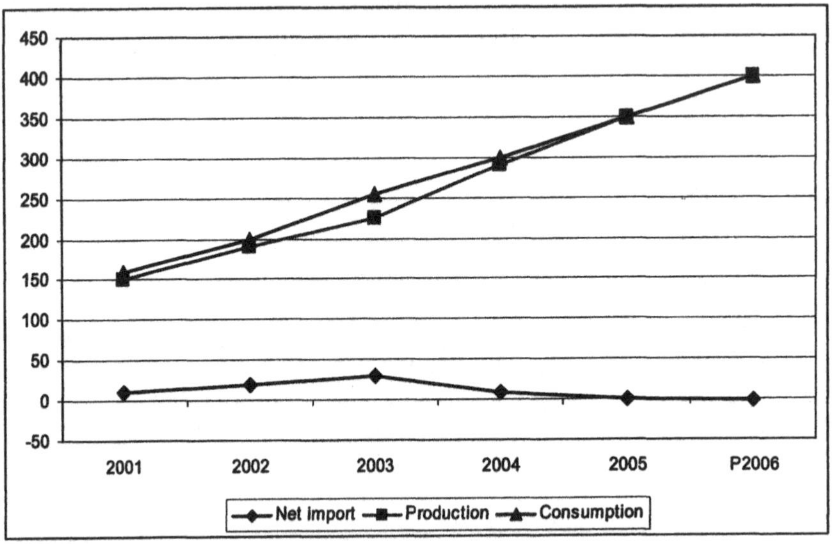

Source: Metla Bulletin

This the level of steel import, after an increase until 2003, started to fall to finally reach the equilibrium of the steel balance in 2005.
The major part of steel import is the coils: 22.48 MT were imported in 2005 out of a total of 26.75 MT.

Fig. 18 - Imports by product in 2005

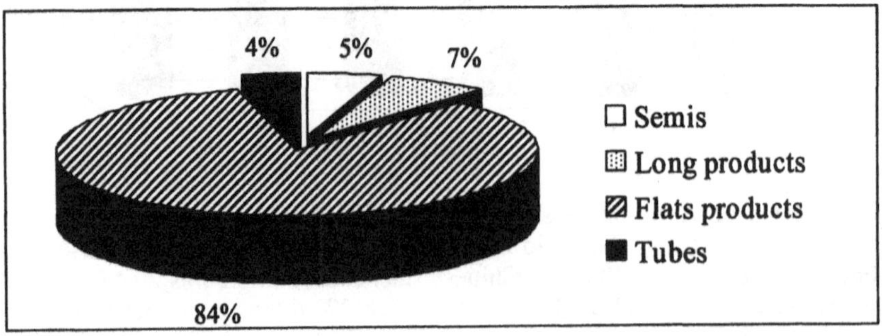

Exports were low until 2002 but started to rise in 2003. The largest growth was registered in 2004 when an increase of 145% was recorded. In 2005, exports continued to rise but slowly at 27.3 MT, raising China to the world's fourth largest exporter behind EU25, Japan and Russia. Sales were principally oriented to east and south Asia.

Fig. 19 - Exports by product in 2005

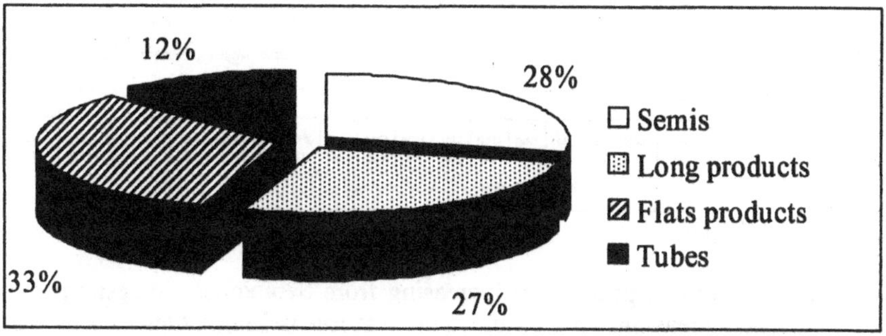

The global deficit was then limited to 1 million tons in 2005 according to these data. It is even planned to note a slight surplus for 2006, it will be almost 10 MT exported.
There is a contradiction between the reality in the Chinese steel industry and the goals of the Chinese government's restructuring strategy (developed hereunder). This affirms that exporting is not a target for China because the capacities and production will fit the internal demand. It is not the case at this time and the marginal tones (important regarding the volume of Chinese steel production) exported to the rest of the world could destabilize the steel market.

Prices
Prices of steel sold in China fell from March 2005 onwards. The composite index of the steel price has decreased from 32% from March to December 2005. The price for coils has decreased from 39% in the same period.
During this period, the gap between domestic prices and international prices increased, particularly during the second semester.

Fig. 20 - Domestic prices evolution for hot rolled

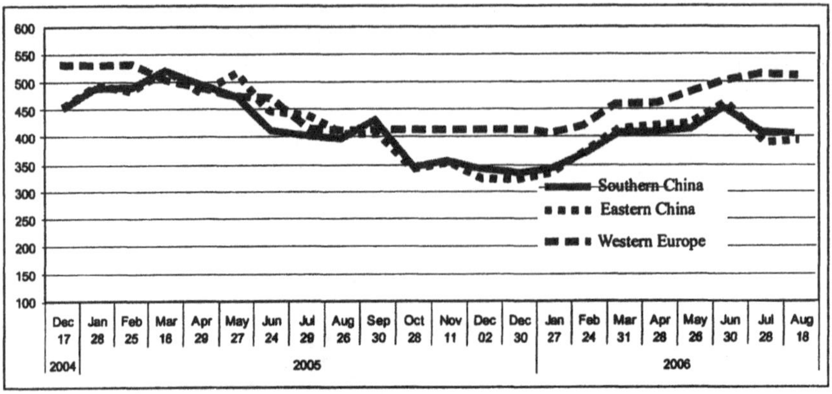

Source: Metal Bulletin

Whereas international prices kept increasing from September, domestic prices in China kept decreasing. This double move was true in long products as well as in flats products, thus was stronger in the latter.
Chinese domestic prices have been increasing after February 2006.

Fig. 21 - Hot rolled coil domestic price evolution (Yuan/ton, price maxi)

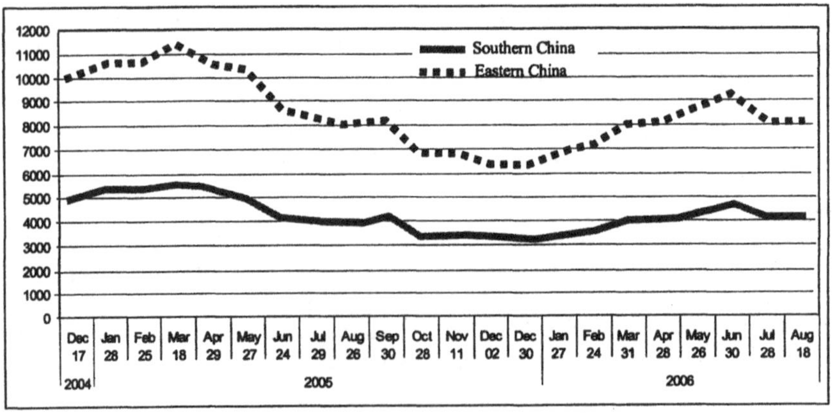

Source: Metal Bullettin

Investments in the Chinese steel industry
Investments in steel plants in China reached 228 billion RMB in 2005, which is an increase of 27.5% in one year. The investments represent around 3% of all fixed asset investments in the country. This ratio remained stable over the past few years.

Tab. 24 - Investment in China's iron and steel industry 2001-2005

	2001	2002	2003	2004	2005
Billion Yuan	48.3	70.4	132.9	191.3	228.1

The production capacities are currently estimated at 414 million tons, which would show a utilization ratio of 85%.
The increase expected considering the current building projects and the future projects would enable China to reach a global production capacity for crude steel of 439 million tons by the end of 2006. Without betting on the implementation of the projects on time, it is possible to underline the weakness of this figure compared to the immediate past when capacity increases have been much more spectacular.

Several elements for interpretation
When prices fell in spring 2005, some diagnosed the effects of the over production resulting from production over capacities in steel. China would then have left the virtuous circle of the steel growth, which had seen it producing from 200 to 350 million tons over a few years.
Since 2002 it is perfectly clear that the world steel industry did not have the production capacities necessary to satisfy the increasing Chinese demand for steel products.The implementation of steel production capacities in a very short time is without any doubt one of the most remarkable industrial event in the world economic history of the past years. Moreover, it has been mostly done without foreign investments and without western technology. For this reason, the Chinese steel industry is characterized by a strongly self centred development.This being said, have these phenomena typical from overproduction occurred in 2005 and 2006 in the Chinese steel industry?
We do not think so, for several reasons:
- The stocks of the producers remained under a one-month delay for delivery
- Prices did fall but they are increasing today

Even if profits have decreased, companies producing steel are far from losing money. Imports have remained high and if exports have increased in 2005, they have not led to many complaints by others producers. In fact, the only relevant indicator to judge the level of under or over production in the steel industry is the utilization ratio of the production capacities.When it is under 80%, there is a risk of over capacity and if it is over 90%, then it shows a tension linked to under capacity.

Tab. 25 - Average production capacity utilization rates*

	2001	2002	2003	2004	2005
China	88,24%	92,35%	84,28%	83,18%	85,12%

**year end production capacity*

The rates for the Chinese steel industry obviously are in agreement with international standards. These utilization ratios are even higher and over 90% considering the average capacities of the year and not the ones of the end of the year. Obviously the question remains to know which plants are really included in these statistic data. But the Chinese steel production growth is not limitless. Beyond the needs in the country in terms of housing, of heavy equipments and of durable consumer goods, the risk is to be confronted with economic and strategic difficulties to supply the production with raw material. But before considering the issue of raw materials and of their availability for the Chinese steel industry, let us see how national authorities implement regulation elements to manage this production and production capacity surplus if needed.

1.2 Chinese government policy

The new steel industrial policy in 2005

On April 13th 2005, the Council of State decided that exports of energy-intensive, polluting and resource-intensive products had to be under supervision. For a few years, investments in the steel, the building, the aluminium and the coke industries have been increasing. Subsidies for exports have been either cancelled or decreased from May 1st 2005. This applies to zinc, tungsten, tin and antimony. A 5% tax on exports is implemented for aluminium and iron alloys. It is now forbidden to trade rare metals, alumina and iron alloys.

As for steel, three types of problems have to be solved:

- The structure of the Chinese steel industry, which is characterized by a low concentration of companies, and by the irrational display of production, which leads to a poor quality of products and to an insufficient innovation capacity.
- No efficient production conditions in the fields of energy, environment protection and transport.
- Very hard working conditions combined with small production units often equipped with obsolete tools.

So the Chinese government promotes a policy favouring:

- The improvement quality in steel products to enable China to be among powerful countries in the field of iron and steel transformation.
- The restructuring of the national steel industry.
- The most efficient units in terms of energy and very respectful to the environment.
- Innovation to promote international competitiveness.
- Restructuring measures.

This new policy was issued in July 2005 by the National Development and Reform Commission (NDRC). The Chinese steel industry restructuring would consist of the removal of around 100 million tons for pig iron capacities and of around 55 million tons for steel production capacities. At the same time, an increased monitoring of investments has been implemented.To give a threshold NDRC plans to shut all the equipments that are less than 300 cubic meters for blast furnaces and 20 tons for converters.

Chinese policy face:
- Over-capacity caused by irrational investments
- Lack of industrial concentration
- Imbalanced product mix
- Poor industrial lay out
- Waste of natural resources and energy, pollutant emissions

For environmental protection and effective utilization of natural resources, consumption of energy and water per crude steel ton should fall from 0.76 tons of coal equivalent and 12 tons of water in 2005 and to 0.73 and 8 tons respectively in 2010 and 0.7 and 6 tons in 2020. In addition steelmakers will develop power generation systems to reuse waste heat and energy at their plant. A few months later, the same NDRC said that progress to shut small and inefficient iron and steel capacity has been unsatisfactory.The Chinese national authority identified two main factors:
- Opposition of the local authorities.
- Rise of the steel prices that encourage producers to invest to increase capacity beyond the threshold.
- The question of over capacity.
- To face over-capacity, the government has taken four core measures:
  a. Imposing a 400 MT limit of overall steelmaking capacity
  b. Elimination of obsolete capacities
  c. Application of new standards for the level of the sector's facilities
  d. Restrictions on new mill projects

According to CISA, China has 414 MT/year of crude making capacity at the end of 2005 and it would increase to 439 MT/year in 2006. The NDRC estimates that 70 MT/year of new capacity is currently under construction and 80MT/year is planned in the sector. In counterpart, NDRC want to force the shut down of the 98.8 MT/year of small blast furnaces and the 55 MT/year of small crude steel making facilities.This last figure can rise at 100 MT/year if the threshold is 30 tones or less for the converters.

Tab. 26 - New standards for the steel sector's facility level (summary)

	New standards	New standards in deep water ports
Coke ovens	6 m high and above	
Blast furnaces	1000 m2 and above	300 m2 and above
Converters	120 T and above	200 T and above
Electric Arc furnaces	70 T and above	
Annual crude steel production		8 MT and above

Source: the new steel policy in OCDE

Information about the social consequences of restructuring is not yet available and is therefore difficult to present.

The workforce in the Chinese steel industry

Tab. 27 - Employees in the chinese rion and steel sectors in 2004

Total number of workers	**1714326**
Blue collar	1037379
White collar	676947
Gender	
Female	442287
Male	1272039
Contract	
Staff and workers contract	1710136
Long contract	1540468
Short contract	169668
Other industry workers	4190

The Chinese steel industry dispersion

In 2005, China has:

- 983 pig iron producers
- 498 crude steel producers
- 2267 producers of steel products

Fig. 22 - Their geographical distribution, even if it is unequal, covers between 70 and 80% of the territory

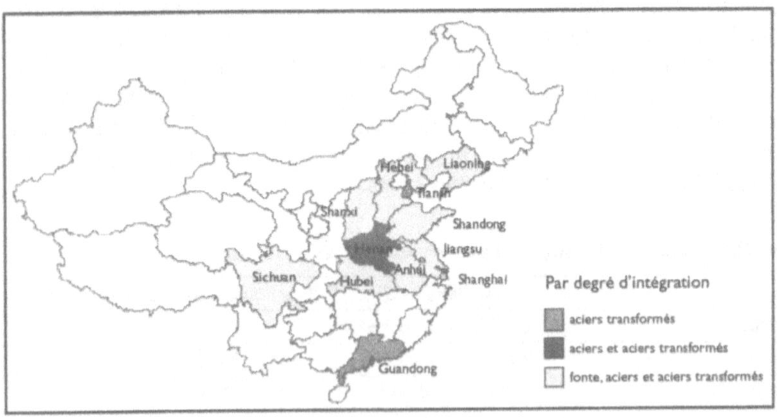

In 2005, the average production per unit was:

- 336 000 tons for pig-iron
- 702 000 tons for steel
- 164 000 tons for steel transformation

More than three quarters of the crude steel production units have produced less than 500 KT in 2005, i.e. only 7,43% of the whole production. As for pig-iron

producers, around 90% of the companies produce less than 500 KT and represent less than 17% of all production. Finally, the ratios are of 94,6% and 25% respectively for steel transformation actors.
Another indicator of the very important dispersion of the Chinese steel industry is the fact that only 17 companies produce more than 5 million tons. Nevertheless the first steps of company concentration have already been taken.
Whereas in 1995 there was no steel company producing 10 million tons or more, there were 8 in 2005. In the same way, there was only one company producing between 3 and 5 million tons, and in 2005, there are 16. According to the steel policy, China plans to establish two to three steel conglomerates with an annual production of over 30 million tons each by 2010 through consolidation, and several others with an output of over 10 million tons. It also expects the top 10 domestic
steel companies will control more than 50% of the country's total output by then, as well as over 70% by 2020.

Concentration and mergers in 2005 and 2006
China seems to gradually getting involved in company concentration in the world steel industry by carrying out a Chinese producer's merger process. The reason for this merger process is the statement that the steel producers do not have much weight faced with their suppliers as well as with their clients. In 2005, the 10 first companies represented less than 40% of the world production. Upstream of the industrial sector, the three main companies exploiting iron ore already represented more than 70% of the whole trade. Lower down, 6 companies in the car industry represented more than 70% of the sales in this sector. A few days later, Wuhan Steel signed an alliance agreement with Hubei Sasac, a shareholder of Echeng Steel. In April, Guizhou Sasac transferred 85% of the Shuigang shares to Shougang. On 24 May 2005, Wuhan steel signed a letter of intention with the government of Guangxi to bring it together with Liuzhou Steel.
Main acquisitions:

Tab. 28 -

Purchaser	Company	New company	Date of Merger	Production* (MT)
Baosteel	Baotou, Bayt, Maan-shan, Jinan, Handan, Shaoguan			22,7
Anshan	Benoci	Anben	Aug 2005	18,4
Tangshan	Chengang	New Tangshan Iron and Steel	Nov 2005	16,1
Wuhan	Liushou	Wugang Ljugang	Dec 2005	17,4
Lalwu Steel	Jlnan Iron and Steel	Shangdong Iron and Steel	Announced for Oct 2006	20,7

**Based on 2005 data*

In November, Baosteel and Wuhan Steel signed a strategic cooperation agreement as Wuhan Steel had already done with Anshan Steel. During the same month, a restructuring agreement of the new Tonggang was signed by Tonggang Jilin Ferroalloys group and Jianlong.The concrete results of this policy are not evident to observe. First if we assess an increase of mergers within the steel industry during last month, this phenomenon seems to be a correction after a greatest dispersion in the last years.

Tab. 29 - The 10 first producers weight in Chinese steel production

	2001	2002	2004	2005
10 first producers (1)	68,67	76,66	94,57	117,51
Chinese steel production (2)	152,1	179,7	270,1	349,9
Ratio (1)/(2)	45,1%	42,7%	35,0%	33,6%

The weight of the first ten steel producers has decreased during the last growth of Chinese steel output.The reason is that the fastest organic growth came from mid sized producers such as Beitai Steel, Kunming Steel and Qingdao, whose output grew by more than 50% between 2004 and 2005. In 2005, 8 steelmakers produced more than 10MT.

The question of foreign investment

The estimated foreign direct investment (FDI) in steel is 1.4% of the total FDI in China. From 1994 to 2004, 5.95 billions US$ have been invested. Japan and Taiwan mainly invest in small case projects, which are below 50 MUS$.They mainly invest in downstream industries.After China's entry to the WTO, the scale of FDI is still expanding.The major steel companies invested in large scale projects, value over 100MUS$. They aimed to satisfy the Chinese domestic demand on the trade markets like automotive and household appliances.These investments are joint ventures linked to the transfer of technologies.

Tab. 30 - Joint ventures

Foreign companies	China's compagnies	Product
Nippon Steel, Arcelor	Baosteel	CR
TKS	Baosteel	STS
TKS	Anshan Steel	Galva
JFE	Guangzhou Steel	Galva
Posco	Benxi Steel	CR

Due to the rapid development of the downstream industries and the technology transfers associated, the encouraged projects by the government in steel industry have in 2005 been limited at only production of direct and fusion direct iron (DRI).

2005 started with the announcement by the Chinese producer Valin of the signature of a contract transferring shares in favour of Mittal Steel. More precisely, this announcement was made on 14 January 2005. In the same year, Arcelor announced the acquisition of a part of capital of Laiwu but for the moment it is not confirmed and some analysts think that foreign direct investment with a majority of control in steel will not be permitted in the next years.

1.3 Steel industry: a strategic industry in China

The limitations encountered by foreign investment in steel are in relation with the strategies of this industry where the great majority of producers are always owned by the public authorities, the state or the province. The Chinese government owns the big steel producers, the subsidies always exist transforming debt to equity, providing subsidized loans, transferring ownership of shares of facilities from one company to another. But beyond the ownership and the strategic character of the steel industry in China we know that the social problems may have a great importance in the management of the sector. Big concentrations of workers in the industrialized provinces where many restructuring measures have already taken place, the social peace must accept some inefficient steelmakers on the markets.

2 The impact and consequences of the Chinese steel boom for the European steel industry

2.1 The surprise of 2004 and its consequences for Europe and beyond

The Chinese steel industry was not a subject in 2003. At that time the birth of Arcelor was always in mind after the announcement of the future shutdown of a lot of continental blast furnaces in Belgium and in France to reduce once more capacities in Europe. The big financial problems in Corus and also internal problems concerning the difficult merger between Dutch and English management was also a major event in 2003. The ECSC treaty died in 2002 and let European steel producers and workers without any institution for industrial dialogue, some discussions were in course but only about R&D.A great part of the North American steel producers went bankrupt and the assistance of the Bush government with protectionist measures and financial support to pay workers pensions and lays off was on line. In brief the western steel industry was very much affected by bad times. The Chinese steel boom has completely changed the situation (Fig. 23).

Only Asia has seen its steel production rise in the last five years.At the beginning of the 2004, the steel world discovers the Chinese economic growth with surprise.The comparison between Europe and China's steel output in the last five years is illustrative. Crude steel production reaches 350 MT in 2005 and more than 400MT is forecasted for 2006. Capacity is estimated by CISA at 414 MT in 2005 and 439MT in 2006. But the first impact of the rise of Chinese steel output has been the rise on raw materials prices.The difficulty to buy steel on the market has offered the possibility for European steel producers to increase the prices. It's the first time that steel prices have not declined during more than three years and have permitted to make the steel producers profitabel (Fig. 24).

Fig. 23 - World steel production by zone

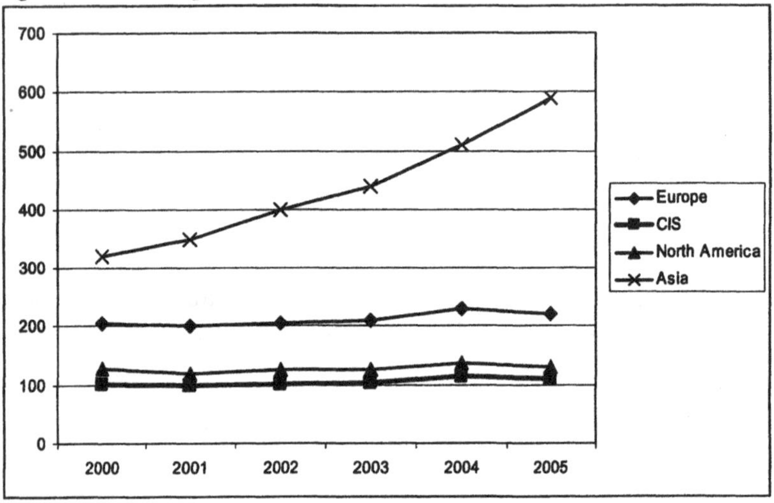

Source: IISI

Fig. 24 - EBITDA 2003-2005 (% of sales)

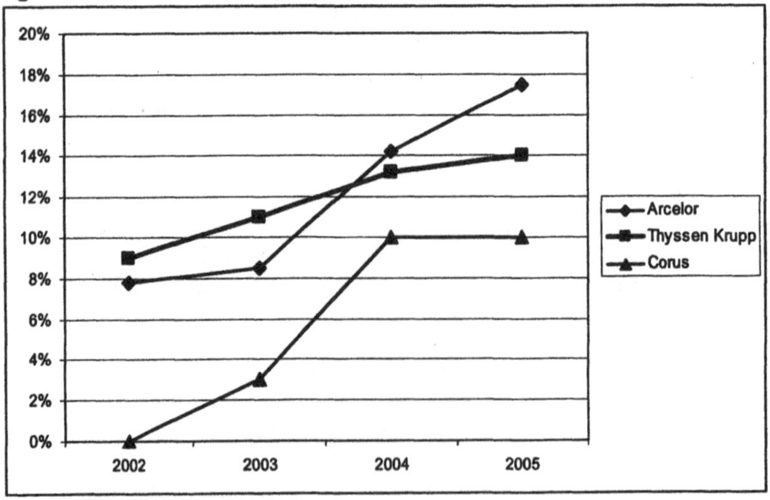

Source: Companies Annual report

The other face of the phenomenon has been the development of the public offers; the first one to Dofasco and the biggest one by Mittal Steel to Arcelor, now the number one steel maker in the world.

2.2 The search of raw materials: a strategic matter for China and for Europe

The Chinese steel boom has woken the steel world first by a big rise of the demand addressed to the exports of steel and another big rise in raw materials; mainly iron ore, coal and scraps. The price of raw materials used in the steel industry has climbed during the last three years. China's companies and government have developed a strategy for the management of raw material seeking security and price control. It is the case in ferrous raw materials even if it is without success at the moment and in non ferrous metals like nickel, which enter in the production of stainless steel.

Iron ore: China's strategy

Iron ore is characterised by an oligopolistic situation where three main producers have the power of the international market to raise the price more than 100% in the last three years.

Fig. 25 - Major iron import countries (MT)

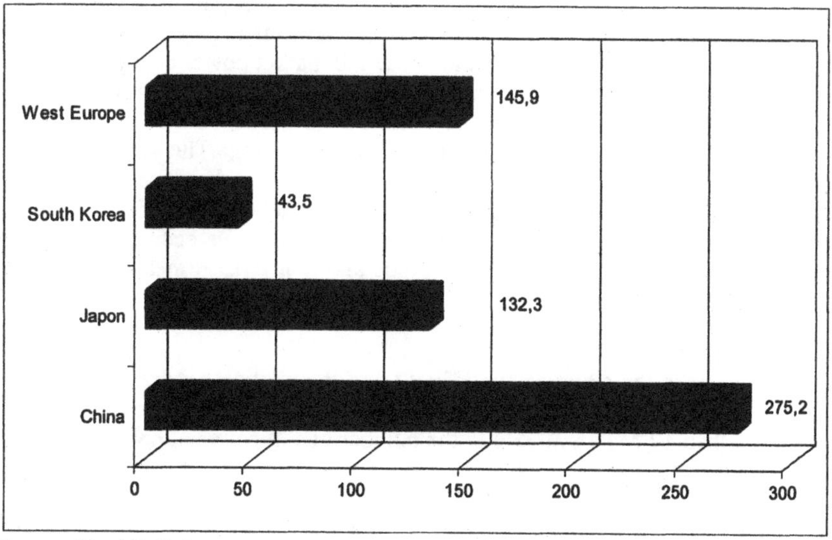

Source: Metal Bulletin

China and Europe are the top importers of iron ore produced in Australia, Brazil and India, and are the third exporter of iron ore. There is no doubt about the availability of iron ore but the question is the security in prices and that iron ore will be available in the next years. Some companies like Mittal Steel have developed an integrated strategy to secure the feed of its industry, so has the Chinese steel industry. Especially in 2006, when the Chinese producers of steel wanted to refuse the rise of iron ore price of 19.5% in comparison with 2005, they observed the poor negotiation power they had to face with the exporters. The strategy of the Chinese steelmakers has been sourcing more iron ore from India, which became the second source after Australia but before Brazil. In the same time they invest in domestic ore mine deposits. The problem is the poor quality of the domestic iron ore, which only have Fe content of 30-33% and many impurities. Moreover safety concerns and environmental legislation will limit the vast number of expansion projects. The access to iron ore will be an important industrial policy subject in the next years between steel producers and countries. Despite a significant increase in mining capacities, the growth of the Chinese demand for raw materials will keep the market fundamentals tight over the next few years. W ith the recent rise in iron ore prices, the mining companies can pay for all expansion plans with only a couple of years of cash flows, the explanation is more than 20 years of underinvestment. So the Chinese strategy will be to invest in foreign mines. For instance in June 2006, the Chinese won in competition with CVRD and Eramet, the right to develop iron ore deposits in Gabon. The Chinese company agreed to build a 260 km long railway, a deep-sea port and a hydroelectric power station as part of the deal. Several large steel producers like Shougang group have made large investments in iron ore deposits all aver the world to erode the market power of the big three. Shougang bought in 2006 a majority stake in Mount Gibson's 5 million T/year magnetite project, while Citic Pacific agreed to invest up to 3 billion USD to buy and develop two iron ore projects in Western Australia.The same company produces iron pellets in Peru (2.5MT/year in 2005).Anshan Steel signed in April 2006 a JV agreement to develop Gindalbie Metals Karara iron ore project.The conclusion is that China seeks raw material for its economic growth, and China is looking to source key ferrous and non ferrous metals to Africa, Latin America and Australia.

The coke situation

Concerning coke the situation is different and the relation between China and Europe is also a strategic matter. The number one exporter of coke is China with more than 12 MT in 2005 and the number one importer of coke is Europe with 10 MT.

The rise of coke prices in 2004 has been a subject of negotiation between the European Commission and the Chinese government with mixed results.At the same time, the growth of the Chinese coke production accelerated to meet the needs of the rising steel production. Exports are still limited by a quotation system, meaning that while coke supply is plentiful within China, it remains scarce on external markets. Most of the new coke ovens are supplied by domestic coking coal.

Tab. 31 - Major importers of Coke (MT)

	2004	2005
Belgium	1,18	0,4
France	2,25	1,52
Germany	7,95	4,93
Italy	1,4	1,06
Netherland	2,96	0,92
Spain	0,17	0,13
United Kingdom	1,05	0,7
West Europe	16,96	9,66
Japan	3	2,7
South Korea	0,7	0,34
Subtotal Asia	3,7	3,04
Total	20,66	12,7

Source: Metal Bulletin

2.3 The question of over capacity

The weight of the Chinese steel industry in the world steel industry transforms the question of the over capacity in China to a global issue.The huge impact of an increase of exports from China is on the same scale as the overcapacity. The over capacity is more than 120MT (10% of the world steel output), which will destabilise the entire sector when it arrives on the market. Some analysts see the net export situation of China since 2006 as the confirmation that is bound to happen. In that perspective, the Chinese over capacity should be a global topic if the Chinese authorities do not succeed in regulating the steel production. It is obvious that this hypothetic situation will have political consequences. Everyone knows that the great majority of the companies are controlled by the government, even if we do not know exactly if it is the central government or provincial governments. Until 2005 and the first part of 2006, we have not seen on the international steel market nor in China the consequences of a structural situation of over capacity. The prices of steel, even if they have decreased in 2005, have permitted to Chinese companies and also other companies in the world to reach a profitable situation. The price development during the last 3 years had been better for all steelmakers without great differences between Chinese internal prices. An equilibrium has been founded despite the big increase of Chinese steel output because the Chinese production was essentially consumed on the internal market. A rise of Chinese exports will be a possibility for Chinese producers but the anti dumping measures used by European and American countries do not always facilitate the possibilities.The perspective of building new capacities in the next two years will not change this situation. The apparent steel use is growing, +136.6 MT of consumption in two years included 84 MT in China (61.5% of the total).

Tab. 32 - Apparent steel use 2005-f 2007 (MT)

	2005	2006	2007	2007/2005
China	315	356	399	84
World except China	698,4	731	751	52,6
World	1013,4	1087	1150	136,6

Source: IISI

This supplementary consumption of steel will be fed by 120-134 MT of new capacities.

Tab. 33 - Capacity addition 2005-2008 (MT)

China	100
Latin America	12
CIS	10
India	10
Asean	2
Total	120-134

Source: OECD steel comitee

The conclusion is that the restructuring plans in China are aimed at reducing but also at replacing the obsolete capacities. So the increase of exports from China is a phenomenon that must have some explanations.The government has provided benefits to exporting companies in the form of tax credits. It also provides a rebate of value added taxes of 11% on steel exports. We can add that the difficulties encountered in restructuring the steel industry may have social reasons. Given that nearly 200 MT of steel products in China is manufactured by small and energy consuming companies who have no way to meet the new industrial policy, a massive restructuring will take place.

2.4 Conclusions

The boom in the Chinese steel industry, which started in 2003, is not a neutral event without consequences on the future of steel in the whole world but is at the contrary a driver of change in the industry.
China is the number on producer and consumer of steel with an industrial policy that wants to

- transform China into a self sufficient country in automobiles
- produce all qualities of steel from the most current to high added value products
- have its main producers owned by public authorities.

These factors have not been a danger, at the contrary; the consequences of the Chinese steel boom have been profitable for the world steel industry. But the surge in 2006 of steel exports is a sign of a phenomenon where the need for good management of the sector is not satisfied. In other terms the difficulties in restructuring encountered by the Chinese authorities should not be exported. In an industry where the cost of work is not an issue, the danger is not relocation like in other industries, but the destabilisation of the market by marginal tonnage which break the equilibrium between consumers and producers. Analysts as well as the Chinese government know that it is necessary to reorganise some steel facilities characterised by inefficiency in energy consumption, bad health conditions and working conditions.The experience of the social dialogue and of steel restructuring in Europe may be a good discussion platform for Chinese officials to solve the social problem they face.At the same time it will let the Chinese government concentrate on the Chinese steel industry and to modernize it.
Europe must be present in raw material discussions to guarantee its access for the manufacturing industry. Europe must control the sources of raw material after a period of low prices and great availability and think about industrial policy in that respect. Beyond the new environmental guidelines published by the Chinese government in 2006, Europe together with China must think about new ways of developing in industry.The Chinese cannot reach the European development model because one planet is not sufficient.The sustainability of future development has not yet a solution.

3 Conclusions

1. Raw materials
   Competition to accede raw material supplies at good prices will be more difficult for the European steel industry. It has already been the case for coke, iron ore and scrap metal.It will strengthen the risks of relocation of one part of the steel industry (hot part).
2. Over capacities
   The development of Chinese over capacities and exports should be a key topic at the European level if the Chinese government does not succeed in its restructuring process. Europe should dispose of tools to prevent this fact and its consequences: One tool is the bilateral contact dialogue groups between the European Commission and China (steel and trade). It has been already the case for coke in 2004.
3. Sustainability
   At the moment the impact of China's growth on the steel industry, especially in Europe, has been very positive because of the price increase and great profits thereof. To meet the interior demand there is no question, but with an export excess we assist at a substitution of obsolete capacities by modern ones. This is permitted by the large support obtained by the China's steel industry

Tab. 34 - Chinese producers with more than 2 MT of crude steel output

	2001	2002	2004	2005
Shanghai Baosteel	19,13	19,48	21,41	22,73
Anshan	8,79	10,07	11,33	11,9
Shougang	8,24	8,17	8,48	10,44
Wuhan	7,08	7,55	9,31	13,05
Benxi	4,9	6,21	5,49	6,51
Maanshan	4,77	5,38	8,03	9,65
Tangshan	3,92	5,07	7,66	10,07
Panzhihua	3,57	5	5,98	6,19
Hualin	4,06	4,91	7,13	8,45
Baotou	4,21	4,82	5,43	7,02
Handan	3,35	4,53	6,8	7,34
Jinan Steel	3,19	3,93	6,87	10,42
Jiangsu Shagang	2,53	3,6	7,55	10,46
Anyang	2,72	3,38	5,24	5,8
Nanjing	1,92	3,2		
Taiyuan	2,84	3,06	4,64	5,39
Laiwu	2,44	2,9	6,58	10,34
Tianjin Tiantie	1,18	2,73	3,45	3,43
Jiangsi Xinyu	1,89	2,32	3,27	4,02
Kunming	2,1	2,27	2,67	3,5
Shaoguan	1,42	2,22	3,46	3,53
Hangzhou	1,68	2,18	2,78	3
Tonghua	1,95	2,09	2,52	2,93
Jiuquan	2,01	2,06	3,7	5,65
Guangzhou	1,9	2,01	3,03	3,03
Total	101,79	119,14	152,81	184,85
Beitai			2,84	4,55
Liuzhou			3,1	4,55
Tangshan Guofeng			4,04	4,54
Nanjing			4,56	4,38
Xuanhua			3,33	3,59
Hebei Jinxi			3,39	3,37
Pingxiang			2,41	3,37
Qingdao			2,25	3,09
Fujian Sangang			2,54	3
Xinjiang Bayi			2,22	2,84
Chongqing			2,64	2,66
Chengde			2,18	2,42
Xingtai			2,16	2,38
Shanxi Haixin			2,47	2,23
Chuanwei			2	2,22
Shijiazhuang			2,08	2,16
Nanchang			1,8	2,14
Lingyuan			2,01	2,11
Tianjin Tangang			1,77	2,08
Chanzhou Zhongtian			1,03	2,02
Shaanxi Longmen			1,22	2,02
Shuigeng			1,83	2,01
Total			53,87	63,73

4. Social aspects and social dialogue
   The steel restructuring process will not happen without social consequences, which today are unknown. The experience of social dialogue in European steel restructuring should help the Chinese government and companies to solve this problem.

5. European Commission
   The participation of trade unions in the discussions between the European Commission and China appears as a necessity (compare bilateral contact dialogue groups).

Chapter 6
Policy recommendations

Caroline Jacobsson

1 Trade Union response

1. Communication between trade unions active in China related activities
   Trade union officers working on China related projects or issues should create a communications network, where they report news and events to each other as well as meet on a regular basis. (Compare with the EMF Eucob@ Correspondents network).
2. Give more attention to European companies active in China
   Trade unions should constantly monitor and put emphasis on individual European companies that have activities (mainly production) in China. The monitoring can lead to an indirect flow of information of state owned enterprises as well since there are emerging joint ventures with SOEs.
3. Evaluate the aftermath of China: what comes next and why?
   Countries and regions around the world are already knocking on the door to follow in the footsteps of the Chinese economic growth. These are amongst others Vietnam,India and Brazil. It is important to monitor the competitive advantages these countries have to be able to adapt accordingly ad not lag behind.
4. What countries are competitors to China?
   Countries competing with business opportunities that can be placed in China are Japan and Korea. This can be a threat to China because these countries have developed economic systems and regulations, which simplifies business relations.
5. Lobby the European Commission to include trade unions in talks
   It is vital for the development of labour issues in China to include trade union representatives in talks with Chinese officials.
6. Chinese opportunities in Europe
   Chinese companies are more and more looking for opportunities in Europe. It is already an established fact in export of cars and joint ventures with European companies in for instance agriculture and services.There is a benefit for European companies and employment is offering the right opportunity to the right Chinese company. Emerging markets is for instance tourism.
7. European Commission Maritime Coordinator
   This person should be petitioned by both the shipbuilding industry and the trade unions to achieve recognition of sector related issues.

Annexes

Interviews

All interviews took place in 2006.

Organisation	Name	Date
EU-China relations		
European Commission, DG Fish	John Richardson	20 April
European Commission, DG Trade	Roelof Plijter	25 April
Delegation of the European Commission to China	Franz Jessen	9 May
European Commission, DG Trade	Alfio Anzalone	19 June
European Commission, DG Trade	Dr. Benedikt Madl	19 June
Shipbuilding*		
BAE Systems (UK)	George Thomson	6 February
BAE Systems (UK)	Emma Harrison	6 February
BAE Systems (UK)	Stuart Gallacher	6 February
BAE Systems (UK)	Paul Moriaty	6 February
Fincantieri (Italy)	Pierfrancesco Tar-tarelli	18 February
JJ Sietas Shipyard	Dr. Christian Bayer	February
Meyer-Werft (Germany)	Hans-Arthur Wilker	3 March
Thyssen Krupp Marine Systems (Germany)	Dr. Klaus Borgschulte	17 March
Thyssen Krupp Marine Systems (Germany)	Reinhard Kuhlmann	17 March
Peene-Werft (Germany)	Mr. Moschell	22 March
Peene-Werft (Germany)	Mr. Kreismer	22 March
Peene-Werft (Germany)	Mr. Kadow	22 March
Volkswerft Stralsund (Germany)	Dr. Christian Bayer	March
University of Warwick	Professor Simon Clarke	14 April
IG Metall Headquarter	Michael Oertel	16 April

Organisation	**Name**	**Date**
IG Metall Headquarter	Claudia Rahmann	16 April
China State Shipbuilding Corporation CSSC	Xiashan Zhang	16 May
China State Shipbuilding Corporation CSSC	Mr. Minshen	16 May
European Commission	Ronald Vopel	25 June
EMF, Sector Committee Shipbuilding	Heino Bade	24 June
VDMA (Marine And Offshore Equipment Industries)	Hauke Schlegel	16 June
HSH Nordbank	Dietmar Hildebrandt	16 June
HSH Nordbank	Dr. Matthias Umlauf	16 June
HSH Nordbank	Ralf Schmidt	16 June
Auto		
IG Metall Ingolstadt (1. Bevollmächtigter)	Johann Horn	10 April
AUDI AG	Wolfgang Förster	10 April
European Automobile Manufactures Association	Michel BRICOUT	26 April
European Commission, DG Enterprise & Industry	Stephanie J MITCHELL	27 April
Fiat Auto Spa	Paolo Arpellino	3 May
Volkswagen (Betriebsrat)	Thomas Rühl	19 May
DaimlerChrysler AG (Management)	Christian Neuhaus	6 June
DaimlerChrysler (Betriebsrat)	Dr.Thomas Metz	6 June
GKN Driveline	Professor Swadhin K. Ghosh	18 July
FIEV French Vehicle Equipment Industries	Jacques Monnet	8 September
TRW Automotive Italia S.p.A.	Luca Zaghi	8 September
Steel**		
European Commission, DG Enterprise	Alberto Canevali	25 April
European commission, DG Trade	Roelof Plijter	25 April
International Iron and Steel Institute (IISI)	Ian Christmas	13 June
European Confederation of iron and Steel Industries	Gordon Moffat	6 July

Organisation	**Name**	**Date**
European Parliament, Committee on industry, research and	Gai Oren	12 July
European Parliament, Economic and Social Committee	Pol Liemans	26 July
European Parliament, Economic and Social Committee	Patrick Venturini	26 July

**Note shipbuilding:*

24 interviews with experts, shipyards directors and trade unionists all over Europe had been carried out in 2004 in the framework of another project (Shipbuilding in Europe), in which the interviewees had been asked about China too. Facts from these interviews have used in this study too.

***Note steel*

Some of the information used in this study has been collected indirectly in the framework of another project including interviews with European steel producers.

References

Asian Intelligence Survey (August 2004), in: *Deutsche Bank Research* 2006

Automotive Industry Action Group (AIAG) ISO/TS 16949:2002 *Implementation Guide*

Baosteel: *Annual report 2005*; "Fact Book" 2005 and 2006

Brewer, Stuart (2006): *China: building for the future*, in: www.dnv.com/publications/dnv_forum/by_subject/classification access at 02.08.2006.

CESA 2006: *Presentation to the EMF China Seminar*, Brussels, 7 November 2006

China Association of Automobile Manufacturers

China Auto Parts online: http://en.auto1688.com.cn/newshow.asp?id=4344

China Statistic Yearbook

Chinese Government's Official Web Portal: www.gov.cn

Clarke S. (2002), *Dalian New Heavy Industry Shipbuilding Co.*, University of Warwick, UK, grey paper

Clarkson Research Services Ltd. (2006): "China Intelligence Monthly",Vol. 1 June 2006

Craig H. (2004), in IG Metall 2004

Deutsche Bank Research (2006): *China 2010+.Welche Faktoren bestimmen die Zukunft Chinas?*, Frankfurt/M 2006, Germany

EUROFER: *Report on the Economic and Steel Market Situation for QIII/2006 and forecast for the Year 2006*

European Commission, DG External relations, http://ec.europa.eu/comm/external_relations/index.htm

China goes global, "Financial Times – Asia edition"

French Government, Ministere de l'économie, des Finances et de l'Industrie – *Adaptation structurelle des équipmentiers de second rang et des PME de la filière automobile* – Rapport final, 2003

GaoY. (2005): *China's shipbuilding industry braces for expansion*, in: www.tdctrade.com/report/indprof/indprof_050907.htm, access at 02.08.2006

Germanischer Lloyd (2005): *Im Riesenreich der Mitte immer vor Ort*, in: "non-stop-Magazin" 04/2005, page 13-20, Hamburg

Hawkins W.R. (2001): *How China plans to Dominate the Shipbuilding Industry*, in: www.americaneconomicalert.org , access at 06.10.2005

Tholen J.,Thorsten L. (2006): *Shipbuilding in Europe – Structure, Employment, Perspectives*; University of Bremen; IAW, www.iaw.uni-bremen.de/downloads/termine/SchiffbauEuropa-paper-en.pdf

ICE (Italian Foreign Trade Institute) and InterChina Consulting – December 2003

IG Metall Bezirk Küste (2004): *Schiffbaunation China – Mit aller Macht nach vorn*, Hamburg

Iron and Steel Statistics Bureau: Articles on the website: www.issb.co.uk; "China watch" accessed on 08.27.2006

KPMG: *China Automotive and Components Market* 2005

Mackey M. (2005*): China's shipping prowess*, in: www.atimes.com/atimes/China/GC19Ad01.html, acces at 28.07.2006.

Malhotra T.C. (2005): *Asia: Shipbuilders to the World*, in: www.scandoil.com/moxie_issue/issue_5-6/2005_5-6/asia-shipbuilders-to-the-world-.shtml; access at 21.05.2006.

Marchisio O., *Siderurgia, cambia tutto: il luogo decisivo ora è la Cina*, in "Liberazione", 06.08.2006

Mattes/Ford and B&D 2005 (Forecast 2007)

Medeiros E.S. et al (2005): *A new direction for China's defense industry*, Santa Monica, RAND Corporation

MEPS: *Statistics* from "Quarter 4/2003", "Quarter 4/2004", "Quarter 4/2005" issues; Articles and statistics on the website http://www.meps.co.uk/

Mercer Management Consulting – 2003

"Metal Bulletin" (specialty revue): Articles from 2006 issues; Articles on the website www.metalbulletin.com

"Metal Bulletin Research": Articles from Steel Markets Monthly

"Metal Bulletin's 4th Far East Steel Conference 2006" in China– Presentations and speeches

Ministry of Commerce of the People's Republic of China, http://english.mofcom.gov.cn/

OECD: Directorate for Science, Technology and Industry – Steel Committee: *Current situation of the Chinese Steel*, Industry 04 April 2006

"OICA Statistics" - http://www.oica.net/htdocs/Main.htm

Relaunching the innovative role of the European automotive industry in the field of technology, organisation and quality of work and The Role of advanced industrial relations Practices

Roth S.: European Metalworker's Federation, Brussels; Chairman of the EMF Automotive Sector Committee:

Schmidt R. (2005): *Chinese shipyards – ample room for improvement*, in: HSH Nordbank (publ.): "Asia Report" 04/2005, page 9-10; Hamburg.

Shipbuilding Action Group Meeting, Placid Harbour, Maryland, USA, 28 August 2006

Shun C. (2005): *Full Steam Ahead*, in: www.bjreview.com.cn/En-2005/05-10-e/10-bus-2.htm, access at 03.07.2006

Stopford M. (2004): *World Shipbuilding 2004*, Advance Press conference SMM 2004, Hamburg 25.5.2004, www.hamburg-messe.de/presse/presse_smm/stopford_en.htm (16 September 2004)

Suzuki M. (2006) *Current State and Outlook of the Chinese Shipbuilding Industry*, Paper, presented to the IMF's Shipbuilding Action Group Meeting, Placid Harbour, Maryland, USA, 28 August 2006

"The Economist", 16 February 2006

The European Commission in Beijing, http://www.delchn.cec.eu.int/

The European Community (2004): *Background report Shipbuilding in China*, Brussels

The European Union Chamber of Commerce in China (European Chamber), http://www.europeanchamber.com.cn/ Thun, E., *Changing Lanes in China: Foreign Direct Investment, Local Governments, and Auto Sector Development*, Cambridge University Press, N.Y., USA, 2006

VDA - German Association of the Automotive Industry database: http://www.vda.de/index_en.html

VSM (2006): Jahresbericht (Annual Report) 2005, Hamburg

VSM (2006): Jahresbericht 2005, Hamburg

Website of Directorate General Trade of the European Commission, http://ec.europa.eu/comm/trade/

Wikipedia open source dictionary, www.wikipedia.com

Yundan G. (2005): *China's shipbuilding industry braces for expansion*, in: www.tdctrade.com/report/indprof/indprof_050907.htm, access at 02 August 2006.

Abbreviations and Vocabulary

ASEAN	The Association of Southeast Asian Nations (ASEAN) is a political and economic organization of countries located in Southeast Asia. ASEAN was formed on August 8, 1967 by the Philippines, Malaysia,Thailand, Indonesia, Singapore, as a display of solidarity against Communist expansion in Vietnam and insurgency within their own borders.
ATTAC	International movement for democratic control of financial markets and their institutions
CEO	Chief executive officer
CESA	Community of European Shipyard's Associations
CGT	Compensated Gross Tonnage *The compensated tonnage of a ship, i.e. the ship's volume adjusted (compensated) by a factor to render the amount of work at the yard equivalent for different types and sizes of ship.*
DG	Directorate General of the European Commission
DWT	Dead Weight Tonnage
EMF	European Metalworkers Federation
EU	European Union
FDI	Foreign Direct Investment
GDP	Gross Domestic Product
GRT	Gross Register Tonnage *Gross Register Tonnage represents the total internal volume of a vessel, with some exemptions for non-productive spaces such as crew quarters; 1 gross register ton is equal to a volume of 100 cubic feet (2.83 m³). Gross register tonnage was replaced by gross tonnage in 1994, under the Tonnage Measurement convention of 1969.*
GT	Gross Tonnage *Since 1994 internationally valid information for the whole measurement of a ship. Gross Tonnage refers to the volume of all ship's enclosed spaces (from keel to funnel) measured to the outside of the hull framing. It is always larger than gross register tonnage (GRT), though by how much depends on the vessel design.*
ICE	Istituto per il Commercio Estero (Italy)
IFA	International Framework Agreement

Immigration	Immigration in the modern sense refers to movement of people from one nation-state to another, where they are not citizens. In the case of economic migration (usually labour migration), differentials in wage rates are prominent. Poor individuals from less developed countries can have far higher standards of living in developed countries, than in their originating countries.
JIT	Just-in-time Just In Time is an inventory strategy implemented to improve the return on investment of a business by reducing in-process inventory and its associated costs.
LNG carrier	Liquefied Natural Gas carrier *Liquid gas tankers must be distinguished between LNG-and LPG-carriers, transporting Liquefield Natural Gas (LNG) or Liquefield Petroleum Gas (LPG). To call LNG, also natural gas, is extracted from pure gas fields. Deep-frozen with possibly – 161 degree centigrade it can be liquefied and shipped in this state.*
Migration	Human migration denotes any movement by humans from one locality to another (migration), often over long distances or in large groups.
MNC	Multinational Company
NGO	Non Governmental Organisation
OECD	Organisation of Economic Cooperation and Development
OEM	Original Equipment Manufacturer
OES	Original Equipment Supplier
Policy	A policy is a plan of action to guide decisions and actions. The term may apply to government, private sector organizations and groups, and individuals.The policy process includes the identification of different alternatives, such as programs or spending priorities, and choosing among them on the basis of the impact they will have. Policies in short can be understood as political, management, financial, and administrative mechanisms arranged to reach explicit goals.
PRC	Peoples' Republic of China
R&D	Research & Development
RMB	Ren Min Bi (People's currency) The official currency of the People's Republic of China
RoPax	Combined Roll-On/Roll-off and passenger ship *with large RoRo decks and limted passenger facilities.*

RoRo	Roll-On/Roll-Off *ship designed for carrying vehicles and wheelbased cargo, which are driven onboard and ashore.*
SOE	State Owned Enterprises
TDW	See DWT
TEU	Twenty foot equivalent unit (container)
TNC	Transnational Corporation
TU	Trade Union
VDMA	Verband Deutscher Maschinen- und Anlagenbau (German Engineering Federation)
VLCC	Very Large Crude Carrier *The VLCC (Very Large Crude Carrier) belongs to the biggest ship types and has a load-carrying capacity of more than 200,000 metric tons. 25 per cent of the fleet are more than 25 years old and have no double hull. 83 per cent of the oil exports from the Persian Gulf are transported with VLCC's. A frequent standard size carries approximately 300,000 metric tons. These ships are approximately 335 metres long and 58 metres wide.*
VSM	Verband für Schiffbau und Meerestechnik (German Association of Shipbuilders and Marine Equipment Manufacturers)
WTO	World Trade Organisation
WWC	World Works Council

Authors

Francesco Garibaldo
Fondazione Istituto per il Lavoro, Bologna, Italy

Daniela Gradinaru
Syndex, Paris, France

Caroline Jacobsson
EMF European Metalworkers' Federation, Brussels, Belgium

Thorsten Ludwig
University of Bremen, IAW Institute Labour and Economy, Germany

Oscar Marchisio
FuzzyNet, Bologna, Italy

Philippe Morvannou
Syndex, Paris, France

Emmanuel Reich
Syndex, Paris, France

Volker Telljohann
Fondazione Istituto per il Lavoro, Bologna, Italy

Jochen Tholen
University of Bremen, IAW Institute Labour and Economy, Germany

Arbeit, Bildung & Gesellschaft
Labour, Education & Society

Herausgegeben von Prof. Dr. György Széll, Prof. Dr. Heinz Sünker,
Dr. Anne Inga Hilsen und Dr. Francesco Garibaldo

Bd. 1 György Széll (ed.): Corporate Social Responsibility in the EU & Japan. 2006.

Bd. 2 Katja Maar: Zum Nutzen und Nichtnutzen der Sozialen Arbeit am exemplarischen Feld der Wohnungslosenhilfe. Eine empirische Studie. 2006.

Bd. 3 Daniela De Ridder: Vom urbanen Sozialraum zur kommunikativen Stadtgesellschaft. 2007.

Bd. 4 Heinz Sünker / Ingrid Miethe (Hrsg.): Bildungspolitik und Bildungsforschung: Herausforderungen und Perspektiven für Gesellschaft und Gewerkschaften in Deutschland. 2007.

Bd. 5 Anja Bastigkeit: Bildungsbiographie und elementarpädagogische Bildungsarbeit. 2007.

Bd. 6 Antônio Inácio Andrioli: Biosoja versus Gensoja. Eine Studie über Technik und Familienlandwirtschaft im nordwestlichen Grenzgebiet des Bundeslandes Rio Grande do Sul (Brasilien). 2007.

Bd. 7 Russell Farnen / Daniel German / Henk Dekker / Christ'l De Landtsheer / Heinz Suenker (eds.): Political Culture, Socialization, Democracy, and Education. Interdisciplinary and Cross-National Perspectives for a New Century. 2007.

Bd. 8 Francesco Garibaldo / Volker Telljohann (eds.): New Forms of Work Organisation and Industrial Relations in Southern Europe. 2007.

Bd. 9 Anne Marie Berg / Olav Eikeland (eds.): Action Research and Organisation Theory. 2007.

Bd. 10 György Széll / Carl-Heinrich Bösling / Ute Széll (eds.): Education, Labour & Science. Perspectives for the 21st Century. 2007.

Bd. 11 Francesco Garibaldo / Philippe Morvannou / Jochen Tholen (eds.): Is China a Risk or an Opportunity for Europe? An Assessment of the Automobile, Steel and Shipbuilding Sectors. 2008.

www.peterlang.de

FSC
www.fsc.org
MIX
Papier | Fördert
gute Waldnutzung
FSC® C083411

Kolibri 360 GmbH
Ferdinand-Jühlke-Straße 7
99095 Erfurt, Deutschland
produktsicherheit@kolibri360.de

Druck:
CPI Druckdienstleistungen GmbH
im Auftrag der
Kolibri 360 GmbH
Ferdinand-Jühlke-Str. 7
99095 Erfurt